Ocular Surface Infection and Antimicrobials

Ocular Surface Infection and Antimicrobials

Editors

**Mark Willcox
Fiona Stapleton
Debarun Dutta**

MDPI • Basel • Beijing • Wuhan • Barcelona • Belgrade • Manchester • Tokyo • Cluj • Tianjin

Editors

Mark Willcox
School of Optometry and
Vision Science
University of New South Wales
Sydney
Australia

Fiona Stapleton
School of Optometry and
Vision Science
University of New South Wales
Sydney
Australia

Debarun Dutta
School of Optometry
Aston University
Birmingham
United Kingdom

Editorial Office
MDPI
St. Alban-Anlage 66
4052 Basel, Switzerland

This is a reprint of articles from the Special Issue published online in the open access journal *Antibiotics* (ISSN 2079-6382) (available at: www.mdpi.com/journal/antibiotics/special_issues/Ocular_Antimicrobials).

For citation purposes, cite each article independently as indicated on the article page online and as indicated below:

LastName, A.A.; LastName, B.B.; LastName, C.C. Article Title. *Journal Name* **Year**, *Volume Number*, Page Range.

ISBN 978-3-0365-6060-1 (Hbk)
ISBN 978-3-0365-6059-5 (PDF)

© 2022 by the authors. Articles in this book are Open Access and distributed under the Creative Commons Attribution (CC BY) license, which allows users to download, copy and build upon published articles, as long as the author and publisher are properly credited, which ensures maximum dissemination and a wider impact of our publications.

The book as a whole is distributed by MDPI under the terms and conditions of the Creative Commons license CC BY-NC-ND.

Contents

About the Editors vii

Preface to "Ocular Surface Infection and Antimicrobials" ix

Debarun Dutta, Fiona Stapleton and Mark Willcox
Ocular Surface Infection and Antimicrobials
Reprinted from: *Antibiotics* **2022**, *11*, 1496, doi:10.3390/antibiotics11111496 1

John E. Romanowski, Shannon V. Nayyar, Eric G. Romanowski, Vishal Jhanji, Robert M. Q. Shanks and Regis P. Kowalski
Speciation and Antibiotic Susceptibilities of Coagulase Negative Staphylococci Isolated from Ocular Infections
Reprinted from: *Antibiotics* **2021**, *10*, 721, doi:10.3390/antibiotics10060721 5

Madeeha Afzal, Ajay Kumar Vijay, Fiona Stapleton and Mark D. P. Willcox
Susceptibility of Ocular *Staphylococcus aureus* to Antibiotics and Multipurpose Disinfecting Solutions
Reprinted from: *Antibiotics* **2021**, *10*, 1203, doi:10.3390/antibiotics10101203 15

Madeeha Afzal, Ajay Kumar Vijay, Fiona Stapleton and Mark D. P. Willcox
Genomics of *Staphylococcus aureus* Strains Isolated from Infectious and Non-Infectious Ocular Conditions
Reprinted from: *Antibiotics* **2022**, *11*, 1011, doi:10.3390/antibiotics11081011 29

Yueh-Ling Chen, Eugene Yu-Chuan Kang, Lung-Kun Yeh, David H. K. Ma, Hsin-Yuan Tan and Hung-Chi Chen et al.
Clinical Features and Molecular Characteristics of Methicillin-Susceptible *Staphylococcus aureus* Ocular Infection in Taiwan
Reprinted from: *Antibiotics* **2021**, *10*, 1445, doi:10.3390/antibiotics10121445 47

Zhi Chen, Jifang Wang, Jun Jiang, Bi Yang and Pauline Cho
The Impact of Antibiotic Usage Guidelines, Developed and Disseminated through Internet, on the Knowledge, Attitude and Prescribing Habits of Orthokeratology Contact Lens Practitioners in China
Reprinted from: *Antibiotics* **2022**, *11*, 179, doi:10.3390/antibiotics11020179 57

Nathaniel S. Harshaw, Nicholas A. Stella, Kara M. Lehner, Eric G. Romanowski, Regis P. Kowalski and Robert M. Q. Shanks
Antibiotics Used in Empiric Treatment of Ocular Infections Trigger the Bacterial Rcs Stress Response System Independent of Antibiotic Susceptibility
Reprinted from: *Antibiotics* **2021**, *10*, 1033, doi:10.3390/antibiotics10091033 67

Kimberly M. Brothers, Stephen A. K. Harvey and Robert M. Q. Shanks
Transcription Factor EepR Is Required for *Serratia marcescens* Host Proinflammatory Response by Corneal Epithelial Cells
Reprinted from: *Antibiotics* **2021**, *10*, 770, doi:10.3390/antibiotics10070770 81

Eric G. Romanowski, Shilpi Gupta, Androulla Pericleous, Daniel E. Kadouri and Robert M. Q. Shanks
Clearance of Gram-Negative Bacterial Pathogens from the Ocular Surface by Predatory Bacteria
Reprinted from: *Antibiotics* **2021**, *10*, 810, doi:10.3390/antibiotics10070810 97

Muhammad Yasir, Debarun Dutta and Mark D. P. Willcox
Enhancement of Antibiofilm Activity of Ciprofloxacin against *Staphylococcus aureus* by Administration of Antimicrobial Peptides
Reprinted from: *Antibiotics* **2021**, *10*, 1159, doi:10.3390/antibiotics10101159 . . . . **107**

Srikanth Dumpati, Shehzad A. Naroo, Sunil Shah and Debarun Dutta
Antimicrobial Efficacy of an Ultraviolet-C Device against Microorganisms Related to Contact Lens Adverse Events
Reprinted from: *Antibiotics* **2022**, *11*, 699, doi:10.3390/antibiotics11050699 . . . . **125**

Parthasarathi Kalaiselvan, Debarun Dutta, Nagaraju Konda, Pravin Krishna Vaddavalli, Savitri Sharma and Fiona Stapleton et al.
Biocompatibility and Comfort during Extended Wear of Mel4 Peptide-Coated Antimicrobial Contact Lenses
Reprinted from: *Antibiotics* **2022**, *11*, 58, doi:10.3390/antibiotics11010058 . . . . **137**

About the Editors

Mark Willcox

Professor Willcox is a medical microbiologist who has worked for many years in the area of infections of medical devices. His laboratory focuses on the development of novel antimicrobials that have applications as antibiotics and disinfectants. He also develops new antimicrobial coatings that can be used for a variety of purposes, including coating medical devices to reduce associated infections. He has taken several of these through to pre-clinical testing, and developed antimicrobial contact lenses up to Phase III clinical trials.

Fiona Stapleton

Professor Stapleton is a clinical scientist with expertise in epidemiology, and experience of basic and clinical research in the fields of corneal infection, dry eyes, and contact-lens-related disease. She was awarded her PhD by City University and Moorfields Eye Hospital in London for her research on the pathogenesis and epidemiology of contact-lens-related disease, and completed a post-doctoral fellowship at University College London.

Debarun Dutta

Dr Dutta's research interests include contact lenses and dry eyes. His dry eye research examines the anterior ocular surface and tear film, tear lipid layer, effect of preservatives and surfactants on ocular comfort. His contact lens research examines contact-lens-related adverse events, particularly infiltrative events, the development of novel antimicrobial agents, antimicrobial peptides, the mechanisms and activity of antimicrobial peptides, and the aetiology of the development of keratitis.

Preface to "Ocular Surface Infection and Antimicrobials"

This collection of articles presents research into ocular surface infections. The collection highlights the importance of staphylococci in these infections, and their susceptibility to common antibiotics. Other papers have outlined how guidelines on the best use of antibiotics can help improve prescribing habits and patient compliance during orthokeratology, and how bacteria respond to antibiotics. Further papers describe new ways of treating or preventing these infections.

Mark Willcox, Fiona Stapleton, and Debarun Dutta
Editors

Editorial

Ocular Surface Infection and Antimicrobials

Debarun Dutta [1,2], Fiona Stapleton [2] and Mark Willcox [2,*]

1. Optometry School Life and Health Sciences, Aston University, Birmingham B4 7ET, UK
2. School of Optometry and Vision Science, University of New South Wales (UNSW), Sydney 2052, Australia
* Correspondence: m.willcox@unsw.edu.au; Tel.: +61-290655394

Infection of the ocular surface can have devastating consequences if not appropriately treated with antimicrobials at an early stage. These infections can lead to blindness through corneal scarring or may lead to the enucleation or evisceration of the globe. Treatment often needs to be fast, empirical and based on disease presentation. However, infection by different microbes (bacteria, fungi, viruses or protozoa) can manifest with similar signs and symptoms, so initial treatment may have to be changed. Furthermore, microbes causing infections are showing increasing resistance to antimicrobials. These delays can worsen outcomes.

This Special Edition was designed to highlight current research in this field, with particular emphasis on the antimicrobial resistance of ocular isolates and developing new ways to prevent or treat ocular infections. Eleven papers have been published in this Special Edition. Their topics range from examining the types of *staphylococci* that cause ocular infections to the development of potentially new ways of treating ocular infections with antimicrobial peptides or predatory bacteria.

Four papers examined the types of *staphylococci* causing ocular infections. The paper by Romanowski et al. [1] reported that the most common species of coagulase-negative *staphylococci* to cause ocular infections was *Staphylococcus epidermidis*, with this species being isolated from ≥84% of endophthalmitis, ≥80% of keratitis and ≥62% of conjunctivitis/blepharitis caused by the coagulase-negative *staphylococci* group. The antibiotic profiles of these coagulase-negative staphylococci suggested that empirical treatments with vancomycin for endophthalmitis and cefazolin or vancomycin for keratitis were appropriate. Afzal et al. [2] found that of 63 *S. aureus* isolates from keratitis in USA or Australia, 87% of all the isolates were multidrug-resistant, and 17% of the isolates from microbial keratitis were extensively drug-resistant. Most Australian strains isolated from keratitis were susceptible to ciprofloxacin, but only 11% of the USA strains were. A follow-up study examining the virulence traits of these strains was published [3]. That study found no significant differences in the frequency of virulence genes between the strains isolated from infections (keratitis or conjunctivitis) compared to those isolated from non-infectious corneal infiltrative events. However, there were differences in the toxin genes produced by the strains isolated from keratitis and conjunctivitis. For example, conjunctivitis strains were more likely to possess genes encoding Panton–Valentine leukocidin. The fourth paper by Chen et al. [4] reported that hospital-acquired methicillin-susceptible *S. aureus* (MSSA) caused a significantly lower rate of keratitis but a higher rate of conjunctivitis than community-acquired MSSA. However, both types of MSSA were highly susceptible to several antibiotics, such as vancomycin and fluoroquinolones. It is worth noting that most centres reserve vancomycin to treat sight-threatening infections due to resistant organisms.

Orthokeratology is used in children to correct myopic refractive error during the day and control the development of myopia (short-sightedness) by modifying the shape of the cornea. These lenses are worn during sleep, which may increase the risk of ocular infection. A study by Chen et al. [5] found that guidelines for the best use of antibiotics during contact lens wear significantly improved prescribing habits, which, in turn, affected the compliance of patients with orthokeratology. Bacteria can respond to antibiotics in

---

Citation: Dutta, D.; Stapleton, F.; Willcox, M. Ocular Surface Infection and Antimicrobials. *Antibiotics* 2022, 11, 1496. https://doi.org/10.3390/antibiotics11111496

Received: 17 October 2022
Accepted: 27 October 2022
Published: 28 October 2022

Publisher's Note: MDPI stays neutral with regard to jurisdictional claims in published maps and institutional affiliations.

Copyright: © 2022 by the authors. Licensee MDPI, Basel, Switzerland. This article is an open access article distributed under the terms and conditions of the Creative Commons Attribution (CC BY) license (https://creativecommons.org/licenses/by/4.0/).

several ways, including activating their stress response processes. The study by Harshaw et al. [6] found that the antibiotics polymyxin B, cefazolin, ceftazidime and vancomycin that target the cell wall or cell membrane of *Serratia marcescens* activate the bacteria's stress response, but antibiotics such as ciprofloxacin that do not act of the cell wall or membrane do not. The stress response may make bacteria tolerant to antibiotics, which may affect the outcomes of infection. A paper by Brothers et al. [7] also studied *S. marcescens*, examining the role of one of its transcription factors, EepR, in keratitis. The study found that mutants lacking *eepR* did not activate a cytokine response in corneal epithelial cells to the same degree as the wild-type strain, implicating EepR in producing pro-inflammatory mediators from *S. marcescens*.

Two papers examined the potential of new therapies for ocular infections. The paper by Romanowski et al. [8] reported on the use of predatory bacteria, *Bdellovibrio bacteriovorus* and *Micavibrio aeruginosavorus* to remove infecting *S. marcescens* or *Pseudomonas aeruginosa* from the eye. These predatory bacteria did not damage the eyes but were also unable to completely remove the infecting bacteria. The authors concluded that the predatory bacteria were no more effective than the normal host defense system at removing infecting bacteria from the eyes. The paper by Yasir et al. [9] examined the ability of antimicrobial peptides to reduce the ability of *S. aureus* to produce biofilms or to reduce the number of preformed biofilms. When used in conjunction with ciprofloxacin, the antimicrobial peptides resulted in substantial reductions in the number of bacteria in preformed biofilms or the ability of *S. aureus* to make biofilms. Importantly, *S. aureus* could not develop resistance to the antimicrobial peptides.

Finally, two papers examined new ways of preventing the microbial colonisation of contact lenses or contact lens cases. Reducing colonisation of lenses or cases might help reduce the incidence of contact lens-related keratitis. Dumpati et al. [10] demonstrated that short (thirty-second) exposure to ultraviolet light of 265 nm (UVC) could significantly reduce the number of *S. aureus* and *P. aeruginosa*, as well as the fungi *Candida albicans* and *Fusarium solani* adherent to contact lenses by up to 3.0 $\log_{10}$. Kalaiselvan et al. [11] reported that an antimicrobial contact lens produced by chemically binding the antimicrobial peptide Mel4 to etafilcon A contact lenses remained biocompatible. There were no significant differences in the clinical responses of an eye wearing the coated or non-coated lenses, nor were there differences in the comfort of the lenses during wear.

In conclusion, this Special Edition highlights important research on ocular infections, which may have consequences for prescribing antibiotics to treat ocular disease, as well as pathogenic mechanisms used by ocular pathogenic bacteria and new potential ways of preventing and treating ocular infections.

Funding: This paper received no external funding.

Conflicts of Interest: The author declares no conflict of interest.

References

1. Romanowski, J.E.; Nayyar, S.V.; Romanowski, E.G.; Jhanji, V.; Shanks, R.M.Q.; Kowalski, R.P. Speciation and Antibiotic Susceptibilities of Coagulase Negative Staphylococci Isolated from Ocular Infections. *Antibiotics* 2021, *10*, 721. [CrossRef] [PubMed]
2. Afzal, M.; Vijay, A.K.; Stapleton, F.; Willcox, M.D.P. Susceptibility of Ocular Staphylococcus aureus to Antibiotics and Multipurpose Disinfecting Solutions. *Antibiotics* 2021, *10*, 1203. [CrossRef] [PubMed]
3. Afzal, M.; Vijay, A.K.; Stapleton, F.; Willcox, M.D.P. Genomics of Staphylococcus aureus Strains Isolated from Infectious and Non-Infectious Ocular Conditions. *Antibiotics* 2022, *11*, 1011. [CrossRef] [PubMed]
4. Chen, Y.L.; Kang, E.Y.; Yeh, L.K.; Ma, D.H.K.; Tan, H.Y.; Chen, H.C.; Hung, K.H.; Huang, Y.C.; Hsiao, C.H. Clinical Features and Molecular Characteristics of Methicillin-Susceptible Staphylococcus aureus Ocular Infection in Taiwan. *Antibiotics* 2021, *10*, 1445. [CrossRef] [PubMed]
5. Chen, Z.; Wang, J.; Jiang, J.; Yang, B.; Cho, P. The Impact of Antibiotic Usage Guidelines, Developed and Disseminated through Internet, on the Knowledge, Attitude and Prescribing Habits of Orthokeratology Contact Lens Practitioners in China. *Antibiotics* 2022, *11*, 179. [CrossRef] [PubMed]

6. Harshaw, N.S.; Stella, N.A.; Lehner, K.M.; Romanowski, E.G.; Kowalski, R.P.; Shanks, R.M.Q. Antibiotics Used in Empiric Treatment of Ocular Infections Trigger the Bacterial Rcs Stress Response System Independent of Antibiotic Susceptibility. *Antibiotics* **2021**, *10*, 1033. [CrossRef] [PubMed]
7. Brothers, K.M.; Harvey, S.A.K.; Shanks, R.M.Q. Transcription Factor EepR Is Required for Serratia marcescens Host Proinflammatory Response by Corneal Epithelial Cells. *Antibiotics* **2021**, *10*, 770. [CrossRef] [PubMed]
8. Romanowski, E.G.; Gupta, S.; Pericleous, A.; Kadouri, D.E.; Shanks, R.M.Q. Clearance of Gram-Negative Bacterial Pathogens from the Ocular Surface by Predatory Bacteria. *Antibiotics* **2021**, *10*, 810. [CrossRef] [PubMed]
9. Yasir, M.; Dutta, D.; Willcox, M.D.P. Enhancement of Antibiofilm Activity of Ciprofloxacin against Staphylococcus aureus by Administration of Antimicrobial Peptides. *Antibiotics* **2021**, *10*, 1159. [CrossRef] [PubMed]
10. Dumpati, S.; Naroo, S.A.; Shah, S.; Dutta, D. Antimicrobial Efficacy of an Ultraviolet-C Device against Microorganisms Related to Contact Lens Adverse Events. *Antibiotics* **2022**, *11*, 699. [CrossRef] [PubMed]
11. Kalaiselvan, P.; Dutta, D.; Konda, N.; Vaddavalli, P.K.; Sharma, S.; Stapleton, F.; Willcox, M.D.P. Biocompatibility and Comfort during Extended Wear of Mel4 Peptide-Coated Antimicrobial Contact Lenses. *Antibiotics* **2022**, *11*, 58. [CrossRef] [PubMed]

Article

Speciation and Antibiotic Susceptibilities of Coagulase Negative Staphylococci Isolated from Ocular Infections

John E. Romanowski, Shannon V. Nayyar, Eric G. Romanowski, Vishal Jhanji, Robert M. Q. Shanks and Regis P. Kowalski *

The Charles T. Campbell Ophthalmic Microbiology Laboratory, Department of Ophthalmology, University of Pittsburgh School of Medicine, Pittsburgh, PA 15213, USA; JER157@pitt.edu (J.E.R.); nayyars@upmc.edu (S.V.N.); romanowskieg@upmc.edu (E.G.R.); jhanjiv@upmc.edu (V.J.); shanksrm@upmc.edu (R.M.Q.S.)
* Correspondence: kowalskirp@upmc.edu; Tel.: +1-412-(647)-7211

Abstract: Coagulase-negative staphylococci (CoNS) are frequently occurring ocular opportunistic pathogens that are not easily identifiable to the species level. The goal of this study was to speciate CoNS and document antibiotic susceptibilities from cases of endophthalmitis ($n = 50$), keratitis ($n = 50$), and conjunctivitis/blepharitis ($n = 50$) for empiric therapy. All 150 isolates of CoNS were speciated using (1) API Staph (biochemical system), (2) Biolog GEN III Microplates (phenotypic substrate system), and (3) DNA sequencing of the *sodA* gene. Disk diffusion antibiotic susceptibilities for topical and intravitreal treatment were determined based on serum standards. CoNS identification to the species level by all three methods indicated that *S. epidermidis* was the predominant species of CoNS isolated from cases of endophthalmitis (84–90%), keratitis (80–86%), and conjunctivitis/blepharitis (62–68%). Identifications indicated different distributions of CoNS species among endophthalmitis (6), keratitis (10), and conjunctivitis/blepharitis (13). Antibiotic susceptibility profiles support empiric treatment of endophthalmitis with vancomycin, and keratitis treatment with cefazolin or vancomycin. There was no clear antibiotic choice for conjunctivitis/blepharitis. *S. epidermidis* was the most frequently found CoNS ocular pathogen, and infection by other CoNS appears to be less specific and random. Antibiotic resistance does not appear to be a serious problem associated with CoNS.

Keywords: coagulase-negative staphylococci; eye infections; endophthalmitis; keratitis; conjunctivitis; blepharitis; API Staph; Biolog; DNA sequencing; *sodA* gene; antibiotic susceptibility

1. Introduction

Coagulase-negative staphylococci (CoNS) are normal inhabitants of the skin and mucous membranes [1]. Coagulase is a protein enzyme that, along with protein A, is bound to and associated with the *Staphylococcus aureus* cell wall. *S. aureus*, by itself, is a serious systemic pathogen of the skin, although there are many species of *Staphylococcus* that do not possess coagulase and are less pathogenic. There are over 45 species of CoNS [2].

Although part of the normal periocular flora [3], CoNS are considered opportunistic pathogens that cause endophthalmitis, keratitis, and conjunctivitis/blepharitis [4]. CoNS endophthalmitis is more common after cataract surgery because of the large load of bacteria inhabiting the eyelid margin [3,5]. CoNS keratitis may be less distinctive because of its association with normal flora, but an abundant number of colonies from corneal specimens obtained for laboratory studies indicate a possible pathogenic etiology [6–9]. CoNS as pathogens of conjunctivitis and blepharitis are not definitively diagnosed, due to a large presence on the eyelids [3], but cases have been described [9–11].

CoNS is generally not identified to the species level from eye cultures, mainly due to expediency. After the identification of *S. aureus* by the presence of coagulase and catalase, there are no practical tests to definitively determine the other staphylococcal species. Biochemicals have been utilized to speciate CoNS without much consistency in identification [6–8]. Pinna et al. speciated 55 CoNS (31 blepharitis, 12 conjunctivitis,

12 keratitis, and no endophthalmitis) into eight species using the commercial kit, API ID 32 Staph (bioMérieux, Paris, France), with more consistency, but there was no comparison with other identification methods [7]. Likewise, without comparison to other methods, Leitch et al. speciated *Staphylococci* from contact lenses using an identification system involving six biochemicals [12]. Their system was able to differentiate *Staphylococcus* into eight species, with a predominance of *S. epidermidis* and *S. capitis/warneri*. Monteiro et al. compared automatic identification (VITEK® 2 system) with conventional methods (biochemicals) and genotypic identification (molecular analysis) of CoNS from blood samples. They found discrepancies within the three methods, but found a better correlation with the conventional methods and genotypic identification. They concluded that the more expensive automated system was more reliable in comparison to phenotypic identification for all bacterial isolates [13].

The first goal of the current study was to speciate CoNS using three methods: (1) API Staph (biochemical system), (2) Biolog GEN III Microplates (phenotypic substrate system), and (3) DNA sequencing of the *sodA* gene, from cases of endophthalmitis, keratitis, and conjunctivitis/blepharitis. The objective was to determine the correlation of *Staphylococcus* species with specific ocular infections. The second goal was to determine the susceptibility patterns of the different species of CoNS, to assure the efficacy of empiric treatment.

2. Results

Table 1 provides the identification of CoNS from endophthalmitis, keratitis, and conjunctivitis/blepharitis using API Staph, Biolog, and DNA sequencing. *S. epidermidis* at 80% (119 of 150) was the most prevalent CoNS species from ocular infections, as determined by the three identification systems. More species of CoNS were noted for conjunctivitis/blepharitis (13) and keratitis (10) than for endophthalmitis (6). Only 16% (24 of 150) of other CoNS isolates were identified with agreement among two or three methods. The Supplementary Information (Table S1) contains the entire data set for the 150 isolates and the results of the three CoNS identification methods.

Table 2 shows the distribution of antibiotics used for the treatment of CoNS from endophthalmitis, keratitis, and conjunctivitis/blepharitis. For endophthalmitis, 100% of CoNS were susceptible to vancomycin and cefazolin. For keratitis, 100% of CoNS were susceptible to vancomycin and 98% were susceptible to cefazolin. For conjunctivitis and blepharitis, CoNS was not highly susceptible (30 to 82%) to any single antibiotic. Cefoxitin was not tested for CoNS conjunctivitis isolates.

Table 1. Identification of coagulase-negative staphylococci (S.) from endophthalmitis, keratitis, and conjunctivitis/blepharitis using API Staph, Biolog, and DNA sequencing.

Isolated from Endophthalmitis	API Staph	Biolog	Sequencing	Correlation of ID Tests	
	n (%)	n (%)	n (%)	3 of 3	2 of 3
S. epidermidis	42 (84)	44 (88)	45 (90)	41	3
S. hominis	3 (6)	1 (2)	1 (2)	0	0
S. lugdunensis	2 (4)	4 (8)	3 (6)	0	3
S. haemolyticus	1 (2)	1 (2)	1 (2)	1	0
S. capitis	1 (2)	0 (0)	0 (0)		
S. aureus	1 (2)	0 (0)	0 (0)		
Isolated from Keratitis					
S. epidermidis	40 (80)	40 (80)	43 (86)	37	4
S. caprae	3 (6)	0 (0)	1 (2)	0	1
S. hominis	2 (4)	3 (6)	0 (0)	0	1
S. warneri	2 (4)	0 (0)	1 (2)	0	1
S. lugdunensis	1 (2)	1 (2)	1 (2)	0	2
S. aureus	1 (2)	0 (0)	1 (2)	0	2
S. capitis	0 (0)	4 (8)	2 (4)		
S. pasteuri	0 (0)	2 (4)	0 (0)		
S. pettenkoferi	0 (0)	0 (0)	1 (2)		
Micrococcus species	1 (2)	0 (0)	0 (0)		

Table 1. Cont.

Isolated from Conjunctivitis/Blepharitis					
S. epidermidis	31 (62)	33 (66)	34 (68)	28	6
S. aureus	6 (12)	0 (0)	0 (0)		
S. haemolyticus	2 (4)	3 (6)	5 (10)	2	0
S. hominis	2 (4)	2 (4)	2 (4)	0	2
S. lugdunensis	2 (4)	4 (8)	2 (4)	1	2
S. warneri	2 (4)	1 (2)	2 (4)	0	1
S. capitis	1 (2)	3 (6)	1 (2)	1	0
S. caprae	1 (2)	1 (2)	2 (4)	0	2
S. chromogenes	1 (2)	0 (0)	0 (0)		
S. cohnii	1 (2)	1 (2)	1 (2)	1	0
S. sciuri	1 (2)	0 (0)	0 (0)		
S. pasteuri	0 (0)	1 (2)	1 (2)		
S. saprophyticus	0 (0)	1 (2)	0 (0)		

Correlation of ID tests is the number of identifications made by the 3 methods; 3 of 3 indicates that all methods had identical species IDs; 2 of 3 indicates that two methods had identical species IDs.

Table 2. Distribution of antibiotic susceptibilities (percent susceptible) for coagulase-negative staphylococci (CoNS). Identification determined by Biolog.

CoNS (Number Identified)										
Endophthalmitis	VA	GM	CIP	OFX	CZ	AMK	CAZ	CC	MXF	FOX
S. epidermidis (44)	100	93.2	47.7	45.5	100	97.7	81.8	84.1	65.9	68.2
S. lugdunensis (4)	100	100	100	100	100	100	100	100	100	100
S. hominis (1)	100	100	100	100	100	100	100	100	100	100
S. haemolyticus (1)	100	100	0	0	100	100	0	0	0	0
Total (50)	**100**	94	52	50	**100**	**98**	82	84	68	70
Keratitis	BAC	VA	GM	CIP	OFX	PB	CZ	TOB	Sulfa	MXF	FOX
S. epidermidis (40)	75	100	87.5	50	50	82.5	97.5	85	82.5	67.5	57.5
S. capitis (4)	100	100	75	75	75	100	100	100	100	100	100
S. hominis (3)	66.7	100	66.7	33.3	33.3	100	100	66.7	66.7	66.7	66.7
S. pasteuri (2)	100	100	100	100	100	100	100	100	100	100	100
S. lugdunensis (1)	0	100	100	0	0	100	100	100	100	0	0
Total (50)	76	**100**	86	52	52	86	**98**	86	86	70	64
Conjunctivitis/ Blepharitis	BAC	ERYT	GM	CIP	OFX	TMP	PB	TOB	Sulfa	MXF
S. epidermidis (34)	79.4	26.5	70.6	47.1	47.1	50	85.3	70.6	79.4	29.4
S. lugdunensis (3)	66.7	0	100	66.7	66.7	33.3	100	100	66.7	66.7
S. hominis (2)	100	0	100	50	50	0	50	100	100	50
S. haemolyticus (3)	66.7	0	33.3	0	0	0	100	0	33.3	0
S. cohnii (1)	0	0	100	100	100	100	100	100	100	100
S. saprophyticus (1)	0	0	100	100	100	100	0	100	100	100
S. capitis (3)	100	100	100	100	100	100	100	100	100	66.7
S. pasteuri (1)	100	100	0	100	100	0	0	0	100	0
S. warneri (1)	100	100	100	0	0	100	100	100	100	0
S. caprae (1)	100	0	100	0	0	0	0	100	100	100
Total (50)	78	30	74	50	50	48	**82**	72	80	36

AMK: amikacin; BAC: bacitracin; CIP: ciprofloxacin; CZ: cefazolin; CAZ: ceftazidime; CC: clindamycin; ERYT: erythromycin; FOX: cefoxitin; GM: gentamicin; MXF: moxifloxacin; OFX: ofloxacin; PB: polymyxin B; Sulfa: sulfisoxazole; TMP: trimethoprim; TOB: tobramycin; VA: vancomycin. Susceptibility was interpreted using the CLSI (Clinical & Laboratory Standards Institute) serum standards. It is assumed that the antibiotic concentrations in the ocular tissue are greater than the concentrations in the blood serum. **BOLD** indicates empiric antibiotics.

Table S2 (Kowalski) is a supplementary table that contains the sequencing data for CoNS identification of ocular isolates.

3. Discussion

The virulence of CoNS as an opportunistic pathogen for ocular infections varies by the diagnosis. There is little doubt that CoNS, at 54% (372 of 684) (Campbell Laboratory data), is the most frequent cause of bacterial endophthalmitis, because the aqueous and vitreous contain no colonizing bacteria [4,14]. The implications of CoNS keratitis and conjunctivitis are supported clinically by the presentation of a large load of CoNS in corneal and conjunctival cultures. There are no distinct classical presentations of CoNS keratitis and conjunctivitis; both an inflamed eyelid margin from a blepharitis patient and a normal eyelid margin wsill present a positive culture for CoNS. Blepharitis is not generally infectious. CoNS is part of the normal flora for the eyelid margin; thus, it is difficult to implicate CoNS as the cause of inflammation. The role of CoNS in clinical blepharitis is based on the ophthalmologist's impression and experience.

Treatment of CoNS ocular infections does not appear to be a therapeutic challenge. Methicillin resistance is not a problem for the treatment of ocular infections because there are effective alternatives for treatment. For endophthalmitis, prevention of CoNS infection is the real dilemma. A battery of topical povidone-iodine, topical antibiotics, and possibly an intracameral injection of antibiotics appears to be effective prophylaxis for most surgical cases [15,16]. Standard treatment of CoNS endophthalmitis is an intravitreal injection of vancomycin (1 mg) (200 µg/mL for a 5 mL vitreous volume). The half-life of vancomycin is 48 h in the inflamed human eye [17]. The present study indicates CoNS to be 100% susceptible to vancomycin.

In general, empiric infectious keratitis, which includes CoNS, is treated topically with fortified cefazolin (50 mg/mL) or vancomycin (20–50 mg/mL), and tobramycin (14 mg/mL) [18]. Fortified vancomycin (100%) and cefazolin (98%) both appear to be effective against CoNS, but both need to be formulated at a pharmacy. Commercially available 0.5% moxifloxacin is also used empirically to treat keratitis [19]. Our in vitro study indicates that moxifloxacin is less effective than vancomycin and cefazolin. The serum standard interpretation of CoNS susceptibility to moxifloxacin was 70% (35 of 50). The 30% resistance may be overreported due to high levels of moxifloxacin in the ocular tissue, which may be effective for treatment [20].

CoNS conjunctivitis is probably, but not definitely, self-limiting. Chronic conjunctival infections have been described with CoNS [10]. Generic antibiotics are generally used for the treatment of conjunctivitis/blepharitis because they are less expensive. Gram-positive topical antibiotics, with a conjunctivitis indication, such as polymyxin B/trimethoprim (82%), sulfacetamide (80%), and gentamicin (74%), may provide better coverage for acute infection. Cefoxitin has not been tested for CoNS conjunctivitis isolates. Beta-lactams are not used for conjunctivitis/blepharitis treatment. Blanco and Núñez indicated that moxifloxacin would provide coverage for both methicillin-susceptible and methicillin-resistant CoNS [21]. In contrast, Thomas et al. reported that the fluoroquinolone anti-infectives demonstrated decreased susceptibility for CoNS, but chloramphenicol (98.4% of 641 isolates) and tetracycline (82.4% of 176 isolates) provided better coverage [22]. It must be noted that if *S. haemolyticus* had not responded to polytrim (polymyxin B and trimethoprim), and was still believed to be a pathogen, the patient may have been placed on vancomycin.

Fortified vancomycin and cefazolin are excessive for CoNS conjunctivitis treatment and are not routinely tested. Blepharitis is generally treated topically with ointments that penetrate and remain longer on the eyelid margins. Bacitracin (78%), erythromycin (30%), and bacitracin/polymyxin B (82%) are sometimes cycled for blepharitis, which is often a chronic condition. The low susceptibility of CoNS to erythromycin (a bacteriostatic antibiotic) may be misleading because erythromycin is a cell-associated antibiotic [23–25]. It is more effective when attached to a cell wall than suspended in a broth. Macrolides can inhibit CoNS biofilm formation [26] and can act as anti-inflammatory agents against the chemotactic factors produced by neutrophils, which lead to eyelid inflammation [26–28].

The original goal of this study was to speciate CoNS and determine species correlations with ocular infections and in vitro susceptibility testing. There does not appear to be a practical and consistent method to definitively speciate CoNS in a timely manner for everyday identification. In contrast to the other two methods, API Staph identified eight CoNS isolates as *Staphylococcus aureus*; only one was identified by DNA sequencing and none by Biolog. All three methods were able to consistently speciate CoNS (80%) to *S. epidermidis*, but only 16% of CoNS were identified as other species. It must be noted that the manual system of Biolog was used instead of the more costly automated system. The manual system was used previously to speciate isolates of Moraxella [29]. Our study did not use MALDI-TOF-MS technology (matrix-assisted laser desorption/ionization-time of flight mass spectrometry), but, in a large-volume microbiology laboratory, CoNS identification to species may be improved using mass spectroscopy [30]. Unfortunately, as a small-volume laboratory, we did not have access to MALDI-TOF-MS, to identify CoNS as an additional comparison. Given the predominance of *S. epidermidis* among isolates and the high levels of susceptibility of CoNS to current antibiotics, a simple coagulase test still appears to be cost-effective and expedient, to distinguish *Staphylococcus aureus* from CoNS. Our study indicates that we need to find consistent methods to identify CoNS species in order to identify correlations with distinct clinical features of ocular disease.

The high concentrations of antibiotics delivered and directed toward ocular tissue are an advantage in the effective treatment of CoNS ocular infections. Antibiotics do not need to travel through the blood system to reach the target tissue. It is a common assumption in ophthalmology that adding an antibiotic directly to the infected site or injecting it into the vitreous provides optimal anti-infective therapy. The need to culture ocular infections and monitor the susceptibility of empiric antibiotics (e.g., vancomycin, cefazolin, moxifloxacin) will ensure future therapeutic success.

4. Materials and Methods

4.1. Coagulase-Negative Staphylococci

CoNS were cultured from patients presenting with endophthalmitis ($n = 50$), keratitis ($n = 50$), and conjunctivitis/blepharitis ($n = 50$) from a single tertiary medical center (University of Pittsburgh Medical Center, Pittsburgh, PA, USA). These cases were submitted for laboratory studies (The Charles T. Campbell Eye Microbiology Laboratory) with specific diagnoses designated on the patient requisition. The isolates were consecutively collected: endophthalmitis (August 2014 to July 2018), keratitis (May 2013 to November 2018), conjunctivitis/blepharitis (May 1998 to September 2018). The location of the culture (e.g., aqueous, vitreous, cornea, conjunctiva, eyelid) supported the diagnosis. Any CoNS growth from an endophthalmitis culture was considered significant as a pathogen, whereas 10 or more colonies on culture from the cornea or conjunctiva were necessary to suspect CoNS keratitis or conjunctivitis. The cut-off of 10 colonies was arbitrary and based on the senior author's experience spanning over 40 years. (RPK). Normal conjunctiva and cornea flora, which includes the ocular surface, has no colonizing bacteria. Any collection of bacteria is generally around 1–4 colonies and probably comes from the eyelid margin. Manipulation by contact lens and administering topical drops could temporarily increase the contamination from the eyelid [12]. These areas are harsh environments for bacterial survival [3]. It must be noted that other reports indicated that 10 or more colonies on the conjunctiva and 100 or more colonies on the eyelid could be significant as pathogens [31,32]. The retrospective study did not require institutional review board/ethics committee approval because direct patient contact and personal information were not involved.

Endophthalmitis cultures were intraocular samples obtained from the aqueous and vitreous of the eye using a syringe and needle. The collected samples (a few drops) were routinely plated on trypticase soy agar supplemented with 5% sheep blood (SBA) (BBL™, Becton, Dickinson and Co., Sparks, MD, USA), aerobic chocolate agar (BBL™), anaerobic chocolate agar (BBL™), Sabouraud dextrose agar supplemented with gentamicin (BBL™), and an enriched thioglycolate broth (BBL™). A few drops of intraocular samples were placed on glass slides for direct examination by Gram and Giemsa stain to observe for microorganisms and cytology. For keratitis, the corneal scraping specimens were cultured directly, using spatulas or jeweler's forceps to place the collected samples on SBA, aerobic chocolate agar, and Sabouraud dextrose agar supplemented with gentamicin. Collected samples were also placed on glass slides for direct examination by Gram and Giemsa stains to observe for microorganisms and cytology. Cultures of the conjunctiva and eyelid were collected with sterile soft-tipped applicators and placed on the same culture media as with keratitis (http://eyemicrobiology.upmc.com/PDFs/SpecimenCollection.pdf) (accessed on 26 February 2021).

As part of a clinical collection of bacteria for laboratory certification studies, bacterial growth on solid media was suspended in broth medium supplemented with 15% glycerol and stored at −80 °C. For this study, these isolates were retrieved by thawing and subculturing on SBA.

4.2. Antibiotic Susceptibility Testing of CoNS

Antibiotics are not only used to treat ocular infections, but also used prophylactically to prevent infections. Ophthalmologists use an array of fluoroquinolones, aminoglycosides, and other classes of antibiotics to treat bacterial infections. In this study, in vitro antibiotic susceptibilities of CoNS were determined using the disk diffusion method [33,34] on Mueller-Hinton II agar (BBL™). There are no susceptibility standards for the topical and intravitreal treatment of ocular infections. Susceptibility was interpreted using the CLSI (Clinical & Laboratory Standards Institute) serum standards; these are used to guide treatment without direct interpretation of susceptibility and resistance. It was assumed that the antibiotic concentrations in the ocular tissue are equal to or greater than the antibiotic concentrations attained in the blood serum.

In our clinical laboratory, routine antibiotic batteries are set up for both Gram-positive and Gram-negative bacteria. Cefoxitin is used to detect methicillin resistance in *Staphylococcus aureus* [33,34]. The antibiotic susceptibilities for CoNS were retrospectively determined from laboratory data used for laboratory certification. Antibiotics tested routinely for the treatment and prophylaxis of endophthalmitis were vancomycin, gentamicin, ciprofloxacin, ofloxacin, cefazolin, amikacin, ceftazidime, clindamycin, moxifloxacin, and cefoxitin. Antibiotics tested routinely for the treatment of keratitis were bacitracin, vancomycin, gentamicin, ciprofloxacin, ofloxacin, polymyxin B, cefazolin, tobramycin, sulfisoxazole, moxifloxacin, and cefoxitin. Antibiotics tested routinely for the treatment of conjunctivitis/blepharitis were bacitracin, erythromycin, gentamicin, ciprofloxacin, ofloxacin, trimethoprim, polymyxin B, tobramycin, sulfisoxazole, and moxifloxacin. Cefoxitin was not tested for CoNS conjunctivitis isolates, since beta-lactam antibiotics are rarely used for treatment.

It was not the intention of this study to recommend treatment or prophylaxis of CoNS ocular infection, but to confirm empiric therapy. Vancomycin is the standard empiric therapy for CoNS endophthalmitis; vancomycin or cefazolin is the standard empiric therapy for CoNS keratitis; conjunctivitis and blepharitis are treated with an array of different therapies based on the ophthalmologist's preference.

4.3. API Staph

The CoNS were retrieved from frozen stocks by sub-culturing on SBA. The CoNS isolates were speciated by API Staph as directed by the package insert (https://www.mediray.co.nz/media/15784/om_biomerieux_test-kits_ot-20500_package_insert-20500.pdf) (accessed on 14 June 2021) (bioMérieux, Chemin de L'Orme, Marcy-L'Étoile, France).

4.4. Biolog

The CoNS were retrieved from frozen stocks by sub-culturing on SBA. Biolog GEN III Microplates (Biolog, Hayward, CA, USA) were used to identify CoNS according to the Biolog methodology (www.biolog.com) (accessed on 14 June 2021). In brief, the medium was inoculated with a CoNS isolate to a turbidity of 90% transmittance and aliquoted to a 96-well microplate at a volume of 0.1 mL per well. The plate was incubated at 34 °C and read manually for color changes at 6 h, 8 h, and 24 h. The tabulated data at each time point were entered into the Biolog Identification Systems Software (OOP 188rG Gen III Database v2.8). Species identification was determined as the most probable as indicated by the software.

4.5. DNA Sequencing

The CoNS were retrieved from frozen stocks by sub-culturing on SBA. The superoxide dismutase gene A (*sod*A) was the target gene for identifying CoNS [35]. This 429-bp-long DNA fragment encodes the manganese-dependent superoxide dismutase in 42 CoNS strains. Chromosomal DNA was obtained using QuickExtract™ DNA solution (Lucigen, Middleton, WI, USA), using the manufacturer's protocol. Sequencing of the *sod*A gene was performed using degenerate primers following the protocol of Poyart et al. [35]. Primers were ordered from Integrated DNA Technologies (Coralville, IA, USA), and Taq DNA polymerase and reagents from New England Biolabs (Ipswich, MA, USA) were used. Sequencing was performed at the University of Pittsburgh Genomic Core facility and analyzed using NCBI BLASTN software [36]. The Supplementary Information (File S1. Kowalski DNA sequence Identification of CoNS) expands the description of CoNS identification by DNA sequencing.

Species were titled if BLASTN results yielded a percent identity over 90% and a high maximum ID score of 240 or greater. The sequences were compared to the other two identification methods for a corresponding match. Samples with poor quality sequence results were re-sequenced. The sequences were either a shorter length than required (~480 bp) or did not match in the BLAST database. Sequences with low similarity scores were sequenced at least twice to confirm the species identification.

Supplementary Materials: The following are available online at https://www.mdpi.com/article/10.3390/antibiotics10060721/s1. File S1: (Kowalski) DNA sequence identification of CoNS. Table S1: (Kowalski) Supplemental data of 150 CoNS isolates and speciation by API Staph, Biolog, and DNA sequencing. Table S2: (Kowalski) Supplemental Sequence Data of 150 CoNS ocular isolates (Case, Identity, Percent ID, Maximum Score, Sequence).

Author Contributions: J.E.R. contributed to the research design; J.E.R. retrieved all the isolates and was responsible for all steps of the DNA sequencing; J.E.R. analyzed the data for CoNS identification; and J.E.R. authored and reviewed the final manuscript. S.V.N. contributed to the research design; S.V.N. retrieved all the isolates and was responsible for all steps of API Staph and Biolog identification; S.V.N. analyzed the data for CoNS identification; and S.V.N. reviewed the final manuscript. E.G.R. contributed to the research design, data analysis, and the review and writing of the final manuscript. V.J. is an ophthalmologist who was our clinical consultant for ocular infections; V.J. was crucial in the review and writing the final manuscript. R.M.Q.S. developed and mentored the DNA sequencing analysis; R.M.Q.S. was vital for teaching J.E.R.; R.M.Q.S. supported the DNA sequencing with materials; R.M.Q.S. was crucial in the review and writing of the final manuscript. R.P.K. was the senior author, main author, provided the isolates through his clinical laboratory, and provided most of the culture media required for the project. R.P.K. had final approval for the experimental design and is the corresponding author for this manuscript. All authors have read and agreed to the published version of the manuscript.

Funding: The Ophthalmology Department has received support from: NIH grants P30-EY08098, the Eye and Ear Foundation of Pittsburgh, Pittsburgh, PA, USA; and unrestricted funds from Research to Prevent Blindness Inc.

Institutional Review Board Statement: The retrospective study did not require approval from an institutional review board/ethics committee.

Conflicts of Interest: The authors have no current "Significant Conflict of Interests" to disclose for the completion of this study as determined by the Office of Research, University of Pittsburgh, Pittsburgh, PA, USA.

References

1. Becker, K.; Skov, R.; von Eiff, C. *Staphylococcus, Micrococcus*, and Other Catalase-Positive Cocci. In *Manual of Clinical Microbiology*, 11th ed.; Jorgensen, J.H., Carroll, K.C., Funke, G., Pfaller, M.A., Landry, M.L., Richter, S.S., Warnock, D.W., Eds.; ASM Press: Washington, DC, USA, 2015; pp. 354–382.
2. Becker, K.; Heilmann, C.; Peters, G. Coagulase-negative staphylococci. *Clin. Microbiol. Rev.* **2014**, *27*, 870–926. [CrossRef]
3. Kowalski, R.P.; Roat, M.I. Normal flora of the human conjunctiva and eyelid. In *Duane's Foundations of Clinical Ophthalmology*; Tasman, W., Jaeger, E.A., Eds.; Lippincott Williams & Wilkins: Philadelphia, PA, USA, 1998; Chapter 41.
4. Kowalski, R.P.; Nayyar, S.V.; Romanowski, E.G.; Shanks, R.M.Q.; Mammen, A.; Dhaliwal, D.K.; Jhanji, V. The prevalence of bacteria, fungi, viruses, and acanthamoeba from 3004 cases of keratitis, endophthalmitis, and conjunctivitis. *Eye Contact Lens* **2020**, *46*, 265–268. [CrossRef]
5. Ormerod, L.D.; Ho, D.D.; Becker, L.E.; Cruise, R.J.; Grohar, H.I.; Paton, B.G.; Frederick, A.R.; Topping, T.M.; Weiter, J.J.; Buzney, S.M.; et al. Endophthalmitis caused by the coagulase-negative staphylococci. 1. Disease spectrum and outcome. *Ophthalmology* **1993**, *100*, 715–723. [CrossRef]
6. Kaliamurthy, J.; Kalavathy, C.M.; Parmar, P.; Thomas, P.A.J. Spectrum of bacterial keratitis at a tertiary eye clinic centre in India. *BioMed Res. Int.* **2013**, 181564.
7. Pinna, A.; Zanetti, S.; Sotgiu, M.; Sechi, L.A.; Fadda, G.; Carta, F. Identification and antibiotic susceptibility of coagulase negative staphylococci isolated in corneal/external infections. *Br. J. Ophthalmol.* **1999**, *83*, 771–773. [CrossRef] [PubMed]
8. Manikandan, P.; Bhaskar, M.; Revathy, R.; John, R.K.; Narendran, K. Speciation of coagulase negative staphylococcus causing bacterial keratitis. *Indian J. Ophthalmol.* **2005**, *53*, 59–60.
9. Foulks, G.N.; Gordon, J.S.; Kowalski, R.P. Bacterial infections of the conjunctiva and cornea. In *Principles and Practices of Ophthalmology*; Albert, D.M., Jakobiec, F.A., Eds.; WB Saunders: Philadelphia, PA, USA, 2000; pp. 893–905.
10. Grasbon, T.; de Kaspar, H.M.; Klauss, V. Coagulase-negative staphylococci in normal and chronically inflamed conjunctiva. *Ophthalmologe* **1995**, *92*, 793–801. [PubMed]
11. Hurley, R. Epidemic conjunctivitis in the newborn associated with coagulase negative staphylococci. *J. Obstet. Gynaecol. Br. Cwith.* **1966**, *73*, 990–992. [CrossRef]
12. Leitch, E.C.; Harmis, N.Y.; Corrigan, K.M.; Willcox, M.D. Identification and enumeration of staphylococci from the eye during soft contact len wear. *Optom. Vis. Sci.* **1998**, *75*, 258–265. [CrossRef]
13. Monteiro, A.C.M.; Fortaleza, C.M.C.B.; Ferreira, A.M.; Calvalcante, R.S.; Mondelli, A.L.; Bagagli, E.; Cunha, M.L.R.S. Comparison of methods for the identification of microorganisms isolated from blood cultures. *Ann. Clin. Microbiol. Antimicrob.* **2016**, *15*, 45–56. [CrossRef] [PubMed]
14. Doft, B.H. Treatment of postcataract extraction endophthalmitis: A summary of the results from the Endophthalmitis Vitrectomy Study. *Arch. Ophthalmol.* **2008**, *126*, 554–556. [CrossRef]
15. Grzybowski, A.J.; Told, R.; Sacu, S.; Bandello, F.; Moisseiev, E.; Loewenstein, A.; Schmidt-Erfurth, U. 2018 Update on intravitreal injections: Euretina expert consensus recommendations. *Ophthalmologica* **2018**, *239*, 181–193. [CrossRef] [PubMed]
16. Grzybowski, A.J.; Kanclerz, P.; Myers, W.G. The use of povidone-iodine in ophthalmology. *Curr. Opin. Ophthalmol* **2018**, *29*, 19–32. [CrossRef]
17. Radhika, M.; Mithal, K.; Bawdekar, A.; Dave, V.; Jindal, A.; Relhan, N.; Albini, T.; Pathengay, A.; Flynn, H.W. Pharmacokinetics of intravitreal antibiotics in endophthalmitis. *J. Ophthalmic. Inflamm. Infect.* **2014**, *4*, 1–9. [CrossRef]
18. Thareja, T.; Kowalski, R.P.; Jhanji, V.; Kamyar, R.; Dhaliwal, D.K. MRSA keratitis and conjunctivitis: What does it mean practically? *Curr. Ophthalmol. Rep.* **2019**, *7*, 110–117. [CrossRef]
19. Durrani, A.; Atta, S.; Bhat, A.K.; Mammen, A.; Dhaliwal, D.K.; Kowalski, R.P.; Jhanji, V. Methicillin-resistant *Staphylococci aureus* keratitis: Initial treatment, risk factors, clinical features, and treatment outcomes. *Am. J. Ophthalmol.* **2020**, *214*, 119–126. [CrossRef]
20. Kowalski, R.P. Perspective: Is Antibiotic Resistance a Problem in the Treatment of Ophthalmic Infections? *Exp. Rev. Ophthalmol.* **2013**, *8*, 119–226. [CrossRef]
21. Blanco, C.; Núñez, M.X. Antibiotic susceptibility of staphylococci isolates from patients with chronic conjunctivitis: Including associated factors and clinical evaluation. *J. Ocul. Pharm. Thera* **2013**, *29*, 803–808. [CrossRef]
22. Thomas, R.K.; Melton, R.; Asbell, P.A. Antibiotic resistance among ocular pathogens: Current trends from the ARMOR surveillance study (2009–2016). *Clin. Optom.* **2019**, *11*, 15–26. [CrossRef]
23. Bosnar, M.; Kelnerić, Ž.; Munić, V.; Eraković, V.; Parnham, M.J. Cellular uptake and efflux of azithromycin, erythromycin, clarithromycin, telithromycin, and cethromycin. *Antimicrob. Agents Chemother.* **2005**, *49*, 2372–2377. [CrossRef] [PubMed]
24. Wingard, J.B.; Romanowski, E.G.; Kowalski, R.P.; Ling, Y.; Bilonick, R.A.; Shanks, R.M.Q. A novel cell-associated protection assay (CAPA) demonstrates the ability of certain antibiotics to protect ocular surface cell lines from subsequent clinical *Staphylococcus aureus* challenge. *Antimicrob. Agents Chemother* **2011**, *55*, 3788–3794. [CrossRef]
25. Wu, E.C.; Kowalski, R.P.; Romanowski, E.G.; Mah, F.S.; Gordon, Y.J.; Shanks, R.M.Q. AzaSite® Inhibits Staphylococcus aureus and Coagulase Negative Staphylococcus Biofilm Formation in vitro. *J. Ocul. Pharm. Ther.* **2010**, *26*, 557–562. [CrossRef]
26. Jain, A.; Sangal, L.; Basal, E.; Kaushal, G.P.; Agarwal, S.K. Anti-inflammatory effects of erythromycin and tetracycline on *Propionibacterium acnes* induced production of chemotactic factors and reactive oxygen species by human neutrophils. *Derm. Online J.* **2002**, *8*, 2.

27. Amsden, G.W. Anti-inflammatory effects of macrolides–an underappreciated benefit in the treatment of community-acquired respiratory tract infections and chronic inflammatory pulmonary conditions? *J. Antimicrob. Chemother.* **2005**, *55*, 10–21. [CrossRef]
28. Desaki, M.; Okazaki, H.; Sunazuku, T.; Omura, S.; Yamamoto, K.; Takizawa, H. Molecular mechanisms of anti-inflammatory action of erythromycin in human bronchial epithelial cells: Possible role in the signaling pathway that regulates nuclear factor-kB activation. *Antimicrob. Agents Chemother.* **2004**, *48*, 1581–1585. [CrossRef] [PubMed]
29. LaCroce, S.; Wilson, M.N.; Romanowski, J.E.; Newman, J.D.; Jhanji, V.; Shanks, R.M.Q.; Kowalski, R.P. *Moraxella nonliquefaciens* and *M. osloensis* are Important *Moraxella* Species that Cause Ocular Infections. *Microorganisms* **2019**, *7*, 163. [CrossRef]
30. Dupont, C.; Sivadon-Tardy, V.; Bille, E.; Dauphin, B.; Beretti, J.L.; Alvarez, A.S.; Degand, N.; Ferroni, A.; Rottman, M.; Herrmann, J.L.; et al. Identification of clinical coagulase-negative staphylococci, isolated in microbiology laboratories, by matrix-assisted laser desorption/ionization-time of flight mass spectrometry and two automated systems. *Clin. Microbiol. Infect* **2010**, *16*, 998–1004. [CrossRef] [PubMed]
31. Szczotka-Flynn, L.; Jiang, Y.; Raghupathy, S.; Bielefeld, R.A.; Garvey, M.T.; Jacobs, M.R.; Kern, J.; Debanne, S.M. Corneal inflammatory events with daily silicone hydrogel lens wear. *Optom. Vis. Sci.* **2014**, *91*, 3–12. [CrossRef] [PubMed]
32. Baleriola-Lucas, C.; Fukada, M.; Willcox, D.P.; Sweeney, D.F.; Holden, B.A. Fibronection concentration in tears of contact lens wearers. *Exp. Eye Res.* **1997**, *64*, 37–43. [CrossRef] [PubMed]
33. CLSI. *Clinical and Laboratory Standards: Performance Standards for Antimicrobial Disk Susceptibility Tests*, 10th ed.; Document M02-A10; Approved Standard; Clinical and Laboratory Standards Institute: Wayne, PA, USA, 2009; Volume 29, Number 1.
34. CLSI. *Performance Standards for Antimicrobial Susceptibility Testing*; Twenty-Third Informational Supplement; CSSI document M100-S23; Clinical and Laboratory Standards Institute: Wayne, PA, USA, 2013.
35. Poyart, C.; Quesne, G.; Boumaila, C.; Trieu-Cuot, P. Rapid and accurate species-level identification of coagulase-negative staphylococci by using the *sodA* gene as a target. *J. Clin. Microbiol.* **2001**, *39*, 4296–4301. [CrossRef]
36. Altschul, S.F.; Gish, W.; Miller, W.; Myers, E.W.; Lipman, D.J. Basic local alignment search tool. *J. Mol. Biol.* **1990**, *215*, 403–410. [CrossRef]

Article

Susceptibility of Ocular *Staphylococcus aureus* to Antibiotics and Multipurpose Disinfecting Solutions

Madeeha Afzal *, Ajay Kumar Vijay, Fiona Stapleton and Mark D. P. Willcox *

School of Optometry and Vision Science, University of New South Wales, Sydney 2052, Australia; v.ajaykumar@unsw.edu.au (A.K.V.); f.stapleton@unsw.edu.au (F.S.)
* Correspondence: m.afzal@unsw.edu.au (M.A.); m.willcox@unsw.edu.au (M.D.P.W.)

Abstract: *Staphylococcus aureus* is a frequent cause of ocular surface infections worldwide. Of these surface infections, those involving the cornea (microbial keratitis) are most sight-threatening. *S. aureus* can also cause conjunctivitis and contact lens-related non-infectious corneal infiltrative events (niCIE). The aim of this study was to determine the rates of resistance of *S. aureus* isolates to antibiotics and disinfecting solutions from these different ocular surface conditions. In total, 63 *S. aureus* strains from the USA and Australia were evaluated; 14 were from niCIE, 26 from conjunctivitis, and 23 from microbial keratitis (MK). The minimum inhibitory (MIC) and minimum bactericidal concentrations (MBC) of all the strains to ciprofloxacin, ceftazidime, oxacillin, gentamicin, vancomycin, chloramphenicol, azithromycin, and polymyxin B were determined. The MIC and MBC of the niCIE strains to contact lens multipurpose disinfectant solutions (MPDSs) was determined. All isolates were susceptible to vancomycin (100%). The susceptibility to other antibiotics decreased in the following order: gentamicin (98%), chloramphenicol (76%), oxacillin (74%), ciprofloxacin (46%), ceftazidime (11%), azithromycin (8%), and polymyxin B (8%). In total, 87% of all the isolates were multidrug resistant and 17% of the isolates from microbial keratitis were extensively drug resistant. The microbial keratitis strains from Australia were usually susceptible to ciprofloxacin (57% vs. 11%; $p = 0.04$) and oxacillin (93% vs. 11%; $p = 0.02$) compared to microbial keratitis isolates from the USA. Microbial keratitis isolates from the USA were less susceptible (55%) to chloramphenicol compared to conjunctivitis strains (95%; $p = 0.01$). Similarly, 75% of conjunctivitis strains from Australia were susceptible to chloramphenicol compared to 14% of microbial keratitis strains ($p = 0.04$). Most (93%) strains isolated from contact lens wearers were killed in 100% MPDS, except *S. aureus* 27. OPTI-FREE PureMoist was the most active MPDS against all strains with 35% of strains having an MIC $\leq 11.36\%$. There was a significant difference in susceptibility between OPTI-FREE PureMoist and Biotrue ($p = 0.02$). *S. aureus* non-infectious CIE strains were more susceptible to antibiotics than conjunctivitis strains and conjunctivitis strains were more susceptible than microbial keratitis strains. Microbial keratitis strains from Australia (isolated between 2006 and 2018) were more susceptible to antibiotics in comparison with microbial keratitis strains from the USA (isolated in 2004). Most of the strains were multidrug-resistant. There was variability in the susceptibility of contact lens isolates to MPDSs with one *S. aureus* strain, *S. aureus* 27, isolated from niCIE, in Australia in 1997 being highly resistant to all four MPDSs and three different types of antibiotics. Knowledge of the rates of resistance to antibiotics in different conditions and regions could help guide treatment of these diseases.

Keywords: *Staphylococcus aureus*; microbial keratitis; conjunctivitis; corneal infiltrative events; antibiotic susceptibility; MPDS susceptibility

Citation: Afzal, M.; Vijay, A.K.; Stapleton, F.; Willcox, M.D.P. Susceptibility of Ocular *Staphylococcus aureus* to Antibiotics and Multipurpose Disinfecting Solutions. *Antibiotics* **2021**, *10*, 1203. https://doi.org/10.3390/antibiotics10101203

Academic Editor: Albert Figueras

Received: 28 August 2021
Accepted: 1 October 2021
Published: 3 October 2021

Publisher's Note: MDPI stays neutral with regard to jurisdictional claims in published maps and institutional affiliations.

Copyright: © 2021 by the authors. Licensee MDPI, Basel, Switzerland. This article is an open access article distributed under the terms and conditions of the Creative Commons Attribution (CC BY) license (https:// creativecommons.org/licenses/by/ 4.0/).

1. Introduction

S. aureus is one of the most common causes of ocular infections worldwide [1]. It has been reported as the most common cause of microbial keratitis (MK), which is a sight-threatening infection of the cornea [2] in Australia [3,4] and the USA, [5,6]. Conjunctival infection (conjunctivitis) is also frequently caused by *S. aureus* [7]. *S. aureus* is also commonly observed in inflammatory adverse reactions associated with contact lens-wearing. These corneal infiltrative events are differentiated into infections or inflammatory conditions; the latter are collectively called non-infectious corneal infiltrative events (niCIE) [8].

Treatment of MK involves the intensive use of topical antibiotics and commonly monotherapy with fluoroquinolones or with the use of fortified antibiotics (for example, a beta lactam such as cefazolin with an aminoglycoside such as tobramycin or gentamicin) [9,10]. Conjunctivitis may be treated by topical application of tetracycline, chloramphenicol, or fluoroquinolones [11]. Conversely non-infectious corneal infiltrative events (niCIEs) are self-limiting and heal upon removal of the contact lens, although prophylactic treatment with topical broad-spectrum antibiotics such as fluoroquinolones, chloramphenicol, and polymyxin B with low dose topical steroids [8] may be used.

S. aureus infections can be difficult to treat because strains may be resistant to multiple antibiotics. *S. aureus* can acquire resistance to virtually every antibiotic that has entered clinical use [12]. Bacteria have developed sophisticated mechanisms of drug resistance to ensure their survival. Resistance to antibiotics can be achieved through multiple biochemical pathways [13] that include modification [14] and destruction of antibiotic molecules [15], decreased antibiotic penetration or increased efflux [16–18], modification or complete replacement, or bypassing of target site [19,20]. The effects of various antibiotics on cytoplasmic peptidoglycan metabolite levels in MRSA were determined and metabolite levels were high in *S. aureus* [21]. Increasing antimicrobial resistance of *S. aureus* has been identified as a public health threat by the World Health Organization [22]. Since emerging in 1961, the incidence and prevalence of methicillin-resistant *S. aureus* (MRSA) in ocular infections has increased dramatically [23,24]. Antibiotic resistance in *S. aureus* can be both inherited and acquired. Inherited resistance [25] includes genes naturally present on chromosomes which confer low membrane permeability, efflux pump expression, and enzymatic inactivation of antibiotics [26]. Acquired resistance includes genetic mutations [27] and horizontal transfer of genes across the strains via mobile genetic elements [28].

Contact lens multipurpose disinfectant solutions (MPDS) are used to disinfect contact lenses when they are not being worn. MPDSs contain disinfectants such as quaternary ammonium compounds or biguanides. *S. aureus* strains which possess *qac* genes can be resistant to disinfectants and are more commonly resistant to antibiotics [22]. As *qac* genes occur alongside genes for antibiotic resistance, there is concern that resistance to disinfectants may increase the spread of antibiotic resistance [29].

There is limited information available on antimicrobial and MPDS susceptibility patterns of clinical isolates of *S. aureus* from Australia in comparison to other countries. The purpose of this study was to investigate the antibiotic and MPDS sensitivities of *S. aureus* isolates from different ocular surface conditions isolated in Australia and the USA.

2. Results
2.1. Antibiotic Susceptibilities

Table 1 summarizes the MIC and MBC of *S. aureus* strains to antibiotics. All isolates were susceptible to vancomycin (100%). The susceptibility to the other antibiotics decreased in the following order: gentamicin (98%), chloramphenicol (76%), oxacillin (74%), ciprofloxacin (46%), ceftazidime (11%), azithromycin (8%), and polymyxin B (8%). Most of the microbial keratitis strains from Australia (isolated between 2006 and 2018) were more commonly susceptible to ciprofloxacin (57%) and oxacillin (93%) compared to microbial keratitis strains from the USA (isolated in 2004) for ciprofloxacin (11%; $p = 0.04$) and oxacillin (11%; $p = 0.02$).

Table 1. Percentage of sensitivity and resistance of *S. aureus* strains from different ocular conditions to antibiotics.

Antibiotic	Microbial Keratitis (n = 23)		Conjunctivitis (n = 26)		niCIE (n = 14)	
	% S	% R	% S	% R	% S	% R
Ciprofloxacin	39.1	60.8	42.3	57.6	71.4	28.5
Ceftazidime	0	100	11.5	88.4	28.5	71.4
Oxacillin	60.8	39.1	76.9	23	92.8	7.1
Gentamicin	95.6	4.3	100	0	100	0
Vancomycin	100	0	100	0	100	0
Chloramphenicol	30.4	69.5	92.3	7.6	78.5	21.4
Azithromycin	0	100	15.3	84.6	7.1	92.8
Polymyxin B	0	100	15.3	84.6	7.1	92.8

Abbreviations: R = resistant; S = susceptible; and niCIE = non-infectious corneal infiltrative events.

Chloramphenicol susceptibility varied by ocular condition and origin of the isolates. In total, 95% of conjunctivitis (isolated in 2006) and 78% of non-infectious CIE strains (isolated between 1995 and 2001) from Australia were susceptible to chloramphenicol. There was a significantly lower rate of susceptibility of microbial keratitis strains from Australia (14%) compared to Australian conjunctivitis strains (95%; $p = 0.04$). There was a similar pattern amongst the USA isolates (isolated in 2004), with 55% of the microbial keratitis strains and 95% of the conjunctivitis strains being sensitive to chloramphenicol. Overall, 30% of microbial keratitis strains from Australia (isolated between 2006 and 2018) and the USA (isolated in 2004) were susceptible to chloramphenicol rather than conjunctivitis (isolated between 2004 and 2006) or non-infectious CIE strains (85%; $p = 0.01$).

Most strains (87%; 55/63) were multidrug-resistant (MDR), which is defined as being resistant to three different classes of antibiotics [22]. Strains 111, 112, and 113 from the USA (microbial keratitis; isolated in 2004) and M43-01 from the Australian (microbial keratitis; isolated in 2018) group (see Table S1, Supplementary Material) were extensively drug-resistant (XDR) strains, which is defined as resistant to almost all antibiotics classes [30]. Strain 32 from Australia (niCIE; isolated in 1997) and strain 46 from the USA (conjunctivitis; isolated in 2004) were susceptible to all antibiotics used. Strains from niCIE (isolated between 1995 and 1999) were more susceptible to antibiotics compared to strains from infections (conjunctivitis + microbial keratitis; isolated between 2004 and 2018). The susceptibility of microbial keratitis strains varied by origin of isolates, with microbial keratitis *S. aureus* strains from the USA being more likely to be MRSA and multidrug-resistant compared to Australian microbial keratitis strains.

2.2. Multipurpose Solution Susceptibility

Isolates from contact lens-related niCIE (isolated between 1995 and 2001) were tested for their susceptibility to the MPDSs. All MPDSs showed good activity against the isolates when used at 100% concentration. After diluting the MPDS, strains were able to grow at different dilutions. Overall, OPTI-FREE PureMoist had the lowest median, namely a median MIC of 5.64% and a median MBC of 11.36%, followed by the Renu Advanced Formula (median MIC of 11.36% and median MBC of 22.72%). Complete RevitaLens OcuTec and Biotrue had similar median MICs of 22.72% and median MBCs of 45.45% (Table 2). There was a significant difference in the MIC between OPTI-FREE PureMoist and Biotrue ($p = 0.02$), where strains were more likely to be resistant to Biotrue. One MDR strain (*S. aureus* 27; isolated in 1997) had a relatively high MIC and MBC, of >90%, compared to Biotrue and Renu Advanced Formula, and moderately high levels for OPTI-FREE PureMoist and Complete RevitaLens Ocutec. The MBCs for all the MPDSs were usually twice the MICs.

Table 2. Minimum inhibitory concentration of MPDSs for *S. aureus* niCIE isolates associated with contact lenses.

S. aureus Strains	OPTI-FREE PureMoist (%)		Renu Advanced Formula (%)		Complete RevitaLens OcuTec (%)		Biotrue (%)	
	MIC	MBC	MIC	MBC	MIC	MBC	MIC	MBC
12	2.84	11.36	2.84	5.64	2.84	5.64	11.36	22.72
20	11.36	22.72	11.36	22.72	22.72	22.72	45.45	90.9
24	5.64	11.36	2.84	11.36	45.45	90.9	11.36	22.72
25	1.42	2.84	1.42	5.64	2.84	5.64	5.64	11.36
26	1.42	5.64	1.42	2.84	5.64	11.36	22.72	45.45
27	22.72	22.72	90.9	90.9	22.72	45.45	90.9	90.9
28	11.36	22.72	11.36	22.72	22.72	45.45	45.45	90.9
29	5.64	11.36	22.72	45.45	22.72	45.45	45.45	90.9
31	11.36	22.72	22.72	45.45	22.72	45.45	5.64	11.36
32	5.64	11.36	22.72	45.45	22.72	45.45	22.72	45.45
33	11.36	22.72	22.72	45.45	22.72	45.45	45.45	90.9
41	5.64	11.36	11.36	45.45	11.36	45.45	22.72	45.45
48	2.84	5.64	2.84	5.64	2.84	5.64	5.64	11.36
117	11.36	22.72	5.64	22.72	11.36	11.36	11.36	22.72

2.3. Antibiotic and MPDS Susceptibility of niCIE Strains

Bacterial strains can be described as susceptible or resistant to an antibiotic; however, there is no such definition for MPDS in the literature. A previous study [31] categorized strains with a MIC greater than 10% as resistant to MPDSs and this classification was used in the current study. While the 10% cut-off used seems arbitrary, it is useful to demonstrate the consequences of improper use of MPDSs during contact lens-wearing as the practice of topping off and reusing MPDSs is a risk factor for infection for contact lens-wearers [32,33]; thus, it is useful to model the consequences of improper use of MPDSs. There was no concordance between antibiotic and MPDS sensitivity (Table 3), thus antibiotic sensitivity was not a good predictor of resistance to MPDSs. One strain (*S. aureus* 27) was resistant to four out of the eight antibiotics and to all MPDSs. Conversely, the strains *S. aureus* 28 and 33, isolated in 1997, were susceptible to six out of the eight antibiotics, while being resistant to all MPDSs (Table 3).

Table 3. Relative susceptibilities of contact lens-related niCIE isolates to antibiotics and MPDSs.

Strains	ANTIBIOTICS								MPDS			
	CIP	CEFT	OXA	GEN	VAN	CHL	AZI	P-B	OPTI	RENU	REV	BIO
12						R	R					R
20		R				R	R	R				
24						R	R		R	R		
25						R	R					
27		R		R		R	R					
28						R	R					
32						R	R			R		
33						R	R					
48		R				R	R					
117	R					R	R			R		
26	R								R		R	
29	R								R			
31	R											
41	R											

No shading indicates that strains were susceptible, and gray indicates they were resistant. Abbreviations: CIP, Ciprofloxacin; CEFT, Ceftazidime; OXA, Oxacillin; GEN, Gentamicin; VAN, Vancomycin; CHL, Chloramphenicol; AZI, Azithromycin; P-B, Polymyxin B; OPTI, OPTI-FREE PureMoist; RENU, Renu Advanced Formula; REV, Complete RevitaLens OcuTec; and BIO, Biotrue.

3. Discussion

This study reports the in vitro susceptibility of ocular strains of *Staphylococcus aureus* from the USA and Australia to commonly used antibiotics and the susceptibility of some strains to contact lens MPDSs. Microbial keratitis strains from Australia (isolated between 2006 and 2018) were more commonly sensitive to fluoroquinolones and oxacillin than the strains from the USA (isolated in 2004). Differences in the antibiotic susceptibility profiles in different geographical populations is not uncommon and may be due to climate [34] or cultural differences [35–38]. One study has shown that widespread over-the-counter supply of antibiotics can underpin high resistance [39] and the ability to access antibiotics in such a way differs between countries.

All strains were susceptible to vancomycin, at 100%, and gentamicin, at 98%. Vancomycin resistance in systemic infections has been reported, [40] however, no resistance has been reported in ocular isolates [4]. Gentamicin is commonly prescribed in *S. aureus* ocular infections, but its susceptibility rates vary [41]. The current results are consistent with other studies from the USA and Australia for *S. aureus* ocular isolates [10,42–44]. The antibiotic susceptibility profile in the current study suggests gentamicin to be the best option to treat *S. aureus* ocular infections in both Australia and the USA, and vancomycin to be reserved to treat isolates that are resistant to other antibiotics.

Overall, less than half (46%; 29/63) of all the strains in the current study were sensitive to ciprofloxacin. Studies from Australia published between 2014 and 2016 reported that 93 to 100% of microbial keratitis isolates were susceptible to ciprofloxacin [45–48]. In contrast, the current study reports increasing resistance of *S. aureus* strains from Australia to ciprofloxacin (66%). The increasing rate of ciprofloxacin resistance in Australian microbial keratitis strains (isolated between 2006 and 2018) is of concern, as fluoroquinolones are the first line of treatment for keratitis in Australia [4]. It would be important to explore this in a larger study. Similarly, the rate of resistance of the USA ocular *S. aureus* isolates (isolated in 2004) to ciprofloxacin in the current study was higher than in Australia. One possible reason is that in Australia, antibiotic use in animals is restricted compared to other countries, including the USA [49], which may account for the low level of resistance of Australian isolates. It is generally believed that bacteria that infect the eye are derived from a general pool of environmental bacteria. Resistant bacteria are transmitted to humans through direct contact with animals [50], through the environment [51], and through food products [52]. The increasing antibiotic resistance worldwide has been attributed to their widespread systemic use, their over-the-counter availability, and their inappropriate use [53] in agriculture and veterinary practices to promote growth and prevent infections in livestock [54,55]. Similarly, in ocular infections, factors such as empirical prescription, short-term exposure, and repeated exposure of antibiotics contributed to the resistance of ocular pathogens [56] and changes in resident ocular flora [57]. A large surveillance study from the USA on the antibiotic resistance among ocular isolates between 2009 and 2016 found that approximately 36% of the ocular *S. aureus* isolates were resistant to ciprofloxacin [58]. An increased proportion of MRSA, from 8.5% to 27.9%, in *S. aureus* isolates collected between 1990 and 2001, has been reported in the USA [59]. MRSA strains are often resistant to fluoroquinolones [58–61]. However, in the current study, only 7% of MRSA strains from Australia (isolated between 2006 and 2018) were ciprofloxacin-resistant, whereas 78% of MRSA strains from the USA (isolated in 2004) were resistant to ciprofloxacin, which is consistent with a previous report from the USA [58]. The mechanism of resistance of ocular MRSA strains resistant to ciprofloxacin is unclear and requires further study.

In the current study, only 11% of *S. aureus* strains were susceptible to ceftazidime and all microbial keratitis strains were resistant to this antibiotic. An increasing rate of resistance of *S. aureus* microbial keratitis isolates to first-generation cephalosporins (cephalothin) over a period of 15 years has been reported [62]. Ceftazidime is generally reported to be active against *S. aureus*, except MRSA strains, but it is less active against *S. aureus* than first and second-generation cephalosporins [63]. Resistance to ceftazidime, a third-generation cephalosporin which can be used to treat MRSA, is horizontally acquired due to β-lactamases or due to alteration and over-expression of the penicillin binding protein [64]. In the current study, the mechanism of resistance may have been different depending on the disease or the country from which the strains were isolated.

In the present study, chloramphenicol remained as a good choice of treatment for conjunctivitis and niCIE caused by *S. aureus*, as 96% and 78% of isolates, respectively, were susceptible. Gram-positive bacteria isolated from microbial keratitis isolates have also been reported regarding low levels of chloramphenicol resistance in the Australian and USA isolates [65,66]. However, the current study findings of the increasing resistance of microbial keratitis strains from Australia [67], isolated between 2006 and 2018, and from the USA (45%), isolated in 2004, are not consistent with these earlier studies and suggest it is a poor choice for treatment of corneal infections. Resistance to chloramphenicol may be inherited [68–70] or acquired [71–73]. The underlying mechanism for the difference in chloramphenicol susceptibility between infectious (MK+ conjunctivitis) and non-infectious ocular conditions requires further investigation.

Most of the *S. aureus* strains in the current study were resistant to azithromycin. Most of the resistant strains were also MRSA, which supports the results of a previous study [10], and most of the strains were resistant to polymyxin B. Polymyxin B is considered a Gram-negative antibiotic that does not diffuse well in mediums and resistance to this antibiotic is characteristic of *S. aureus* [74]. This study supports previous recommendations that Polymyxin B is not a good choice for the treatment of *S. aureus* ocular infections [75].

Only 6% of Australian strains (2/32), isolated between 2006 and 2018, were resistant to oxacillin (i.e., could be classified as MRSA), and conversely, 45% of all the USA strains (14/31), isolated in 2004, were resistant to oxacillin. In the USA, an increase in the proportion of MRSAs among *S. aureus* ocular isolates, specifically from 29.5% in 2000 to 41.6% in 2005, has been reported in a national surveillance study (ARMOR) [10]. The high level of MRSAs among *S. aureus* isolates is of concern as MRSA is believed to cause more severe diseases than methicillin-sensitive *S. aureus* [76]. Further molecular analysis of the geographical variation of MRSA in the USA and of the Australian microbial keratitis and conjunctivitis strains, as well as of community or hospital-acquired MRSA, is required.

The study has demonstrated that niCIE strains of *S. aureus*, isolated between 1995 and 2001, vary in their susceptibilities to MPDSs. Most of the strains were susceptible to all MPDSs when used at 100% concentrations, indicating a good activity of the MPDSs. The most effective MPDS, specifically OPTI-FREE PureMoist, contains two disinfectants, namely Polyquaternium-1 and Aldox. Polyquaternium-1 showed good activity against *S. aureus* when used alone, as Aldox has been shown to do, as well [77]. Renu Advanced was the second most effective MPDS in the current study. It contains three disinfectants, namely alexidine, PAPB, and polyquaternium-1. All these disinfectants have been reported to be effective against bacteria [77–80] and some against their biofilms [81].

Complete RevitaLens, containing alexidine and Polyquaternium, was the third most effective MPDS against *S. aureus* isolates in the present study, but has also been reported to show equal efficacy to OPTI-FREE against *S. aureus* in a previous study [82]. Even though both the disinfectants are effective against *S. aureus* [77,80], dilution of the MPDS decreased its efficacy. Biotrue was the least effective MPDS against *S. aureus* isolates in the current study. Biotrue contains only polyaminopropyl biguanide (PAPB, which is also known as polyhexamethylene biguanide (PHMB). PAPB is active against *S. aureus* [83] but its efficacy is concentration-dependent [84]. One study reported a reduced concentration of PAPB (PHMB) after soaking contact lenses in Biotrue and this lower concentration was associated with its decreased antimicrobial activity against *S. aureus* [84]. The findings of the current study regarding the most to least active MPDSs against *S. aureus* are, in general, in agreement with another study [84].

Resistance to disinfectants can be mediated by the *qac* gene, which can be carried on the same transmissible elements as antibiotic resistance genes [67,84]. While possession of *qac* has been associated with resistance to antibiotics [84], there was no clear phenotypic relationship between antibiotic and MPDS resistance observed in the current study. These strains have not been genotyped previously and exploring whether these strains possess the *qac* gene would help to understand the genotypic relationship between antibiotic and MPDS resistance. Other issues could be addressed in future studies by exploring the biocides in the MPDS as well as their dilutions and effects on MIC individually and in combination with other biocides.

The current study used a convenience sample of strains within the culture collection. All strains from the USA were isolated in 2004. Surveillance studies have shown that the rates of the methicillin resistance of *S. aureus* isolated from keratitis in the USA has not changed from 1997 to 2012 [41]. Overall antibiotic susceptibility has shown little or no change in the resistance patterns of ocular *S. aureus* over the periods of 2009–2013 [85] and 2009–2016 [10]. Similarly, strains isolated from keratitis in Australia between 2005 and 2015 showed little or no change in antibiotic susceptibility to ceftazidime, gentamicin, chloramphenicol, fluoroquinolones, and vancomycin [4]. This panel of antibiotics were used in the current study. The Australian strains used in this study were isolated between 1995 and 2018, with the majority from infection isolated between 2006 and 2018 (17/18). Understanding the susceptibility pattern of these strains could help to reduce the risk of inappropriate antimicrobial prescribing. However, as resistance rates can change over time, future studies should examine strains isolated within matched timeframes. Another issue that could be addressed in future studies is whether the use of combinations of antibiotics can overcome any of the resistance observed.

4. Materials and Methods

4.1. Staphylococcus aureus Isolates

In total, 63 *S. aureus* clinical isolates were evaluated (Table 4). Strains from the Bascom Palmer Institute, Miami (USA), were kindly provided by Dr Darlene Miller, while those from the Prince of Wales Hospital (Australia) were kindly provided by Dr. Monica Lahra. All strains were stored in culture collection at the School of Optometry and Vision Science, UNSW. The identity of the strains was confirmed using the automated identification system VITEK 2 for Gram-positive bacteria (BioMérieux, Baulkham Hills, NSW, Australia) according to the manufacturer's instructions.

4.2. Susceptibility to Antibiotics

The susceptibility of *S. aureus* strains to different antibiotics was assessed according to the standard protocol described by the Clinical and Laboratory Institute [86]. Antibiotics commonly used to treat these ocular conditions in Australia and in the USA were selected for the test panel and antibiotic stock solutions were prepared following the manufacturer's recommendations. Antibiotics were diluted in Mueller-Hinton II broth (cation-adjusted, Becton Dickinson and Company, Franklin Lakes, NJ, USA) in sterile 96-well plates to provide the final concentrations ranging from 5120 µg/mL to 0.25 µg/mL.

Bacterial cells at a final concentration of 1×10^5 CFU/mL were then inoculated into 96-well plates with different dilutions of antibiotics and incubated at 37 °C for 18–24 h. For Oxacillin and Vancomycin MIC, *S. aureus* strains were incubated at 35 °C according to CLSI standards [86]. Growth turbidity was measured using a spectrophotometer (FLUOstar Omega, BMG LABTECH, Ortenberg, Germany) at 660 nm. The MIC was taken as the lowest concentration of an antibiotic with no visible growth. For minimum bactericidal concentration (MBC), viable counts were performed by subculturing the cells onto Mueller-Hinton agar (Becton Dickinson and Company, Franklin Lakes, NJ, USA) at their MIC and at the next two higher dilutions of antibiotics; afterwards, they were incubated at 37 °C for 18–24 h. The MBC was the concentration of antibiotics that showed 99.99% bacterial killing [87,88]. The results were interpreted using breakpoints from the Clinical and Laboratory Standards Institute [86] and the European Committee on Antimicrobial Susceptibility Testing [88]. Both resistant and intermediate resistant strains were considered resistant for the subsequent analyses.

Table 4. *S. aureus* ocular isolates used in the study.

S. aureus Isolates	Origin	Associated Condition	Year of Isolation
106	Bascom Palmer Institute, Miami (USA)	Microbial keratitis (MK)	2004
107			
108			
109			
110			
111			
112			
113			
114			
129			2006
34			1997
M5-01	Prince of Wales Hospital (Australia)		2018
M19-01			
M27-01			
M28-01			
M30-01			
M36-01			
M43-01			
M49-02			
M65-02			
M71-01			
M90-01			
M91-01			

Table 4. Cont.

S. aureus Isolates	Origin	Associated Condition	Year of Isolation
84	Bascom Palmer Institute, Miami (USA)	Conjunctivitis	2004
85			
86			
87			
88			
89			
90			
91			
92			
93			
94			
95			
96			
97			
98			
99			
100			
101			
102			
103			
104			
105			
46	Prince of Wales Hospital (Australia)		2006
134			
136			
140			
12	SOVS, UNSW (Australia)	Contact lens-related non-infectious corneal infiltrative events (niCIE)	1995
20			
24			1996
25			
26			
27			
28			
29			1997
31			
32			
33			
41			1999
48			2001
117			1999

4.3. Susceptibility to Multipurpose Disinfectant Solutions

Susceptibility of the bacterial strains isolated from contact lens-related niCIE to four commercially available MPDSs (Table 5) was assessed. This testing was restricted to these isolates as all other strains were isolated from non-contact lens-wearers. The MPDSs were OPTI-FREE PureMoist (Alcon, Fort Worth, TX, USA), Complete RevitaLens OcuTec (Abbot Medical Optics, Hangzhou, China), and Biotrue and Renu Advanced Formula

(Bausch + Lomb, Rochester, NY, USA; Table 5). MPDS susceptibility was tested using previously published methods [31,85]. In brief, each MPDS was serially diluted in freshly prepared sterile phosphate-buffered saline (NaCl 80 g/L, Na_2HPO_4 11.5 g/L, KCl 2 g/L, and KH_2PO_4 2 g/L, pH = 7.2) to protect the bacteria from pH shock. The serially diluted MPDS (200 µL) was added to wells of a microtiter plate and a 20 µL bacterial suspension was added to achieve a final concentration of 1×10^5 CFU/mL. Positive (PBS + bacteria) and negative controls (undiluted PBS) were used. The plates were incubated at 37 °C for 18–24 h. Growth turbidity was measured using a spectrophotometer (FLUOstar Omega, BMG LABTECH, Germany) at 660 nm. Strains with a MIC of more than 10% MPDS were considered resistant. MBC was the concentration of the MPDS that gave 99.99% (3 log units) bacterial killing [85,89]. The purpose of testing MPDSs outside the stated instruction was to find the MIC of *S. aureus* that caused corneal infiltrative events, as concentrations of disinfectants through topping off or through the reuse of disinfecting solutions have been identified as a risk factor for contact lens-related corneal infections [33]. There is some evidence that this may occur more frequently with certain MPDS products, thus it is not unreasonable to challenge MPDS products in a way that may mimic their use in the community.

Table 5. Multipurpose disinfecting solutions and their active agents.

MPDS	Manufacturer	Disinfectants and Their Concentrations
OPTI-FREE® PureMoist®	Alcon, Fort Worth, TX, USA	Polyquaternium-1, 10 ppm; Aldox, 6 ppm
Complete RevitaLens OcuTec (now sold as ACUVUE™ RevitaLens)	Abbot Medical Optics, Hangzhou, ZJ, China (Johnson and Johnson Vision)	Alexidine dihydrochloride, 1.6 ppm; polyquaternium-1, 3 ppm
Biotrue®	Bausch + Lomb, Rochester, NY, USA	Polyaminopropyl biguanide, 1.3 ppm; polyquaternium-1, 1 ppm
Renu® Advanced Formula		Polyaminopropyl biguanide, 0.5 ppm; polyquaternium-1, 1.5 ppm; alexidine, 2 ppm

4.4. Statistical Analysis

Differences in the frequency of antibiotic susceptibility between infectious (MK+ conjunctivitis) and non-infectious (niCIE) groups from Australia and the USA, and MPDS susceptibility in contact lens-related niCIE strains were only compared using Fisher's exact test (GraphPad prism, 2019, v8.0.2.263). For all analyses, a p-value of <0.05 was considered statistically significant.

5. Conclusions

This study concludes that *S. aureus* strains isolated from microbial keratitis from the USA (isolated in 2004) were more likely to be MRSA and multidrug-resistant compared to Australian microbial keratitis strains (isolated between 2006 and 2018). In addition, microbial keratitis strains from the USA and Australia were less susceptible to antibiotics compared to conjunctivitis (isolated in 2004–2006) and non-infectious CIE strains (isolated between 1995 and 2001). Exploring the genomic resistance mechanisms and possession of virulence traits between infectious (MK+ conjunctivitis) and non-infectious ocular conditions from the USA and Australia may help to understand these susceptibility findings. The findings of this study will help to understand the resistance pattern of ocular *S. aureus* isolates from the USA and Australia, which will further inform treatment options.

Supplementary Materials: The following are available online at https://www.mdpi.com/article/10.3390/antibiotics10101203/s1, Table S1: Details of the MIC and MBC of *S. aureus* strains from different ocular conditions to the antibiotics used in the current study.

Author Contributions: Conceptualization, M.A., M.D.P.W., F.S. and A.K.V.; methodology, M.A., M.D.P.W., F.S. and A.K.V.; writing—original draft preparation, M.A.; writing—review and editing, M.D.P.W., F.S. and A.K.V.; supervision, M.D.P.W., F.S. and A.K.V.; funding acquisition, M.D.P.W. All authors have read and agreed to the published version of the manuscript.

Funding: This research study received no external funding.

Data Availability Statement: Data supporting this study is available in the supplementary material.

Acknowledgments: The authors would like to acknowledge Darlene Miller from the Bascom Palmer Institute, Miami (USA), and Monica Lahra from the Prince of Wales Hospital, Sydney, for providing *S. aureus* MK strains. The study was supported by UNSW Sydney, Australia.

Conflicts of Interest: All the authors declare no conflict of interest.

References

1. Mainous, A.G., 3rd; Hueston, W.J.; Everett, C.J.; Diaz, V.A. Nasal carriage of *Staphylococcus aureus* and methicillin-resistant *S. aureus* in the United States, 2001–2002. *Ann. Fam. Med.* **2006**, *4*, 132–137. [CrossRef]
2. Schaefer, F.; Bruttin, O.; Zografos, L.; Guex-Crosier, Y. Bacterial keratitis: A prospective clinical and microbiological study. *Br. J. Ophthalmol.* **2001**, *85*, 842–847. [CrossRef]
3. Mah, F.S.; Davidson, R.; Holland, E.J.; Hovanesian, J.; John, T.; Kanellopoulos, J.; Shamie, N.; Starr, C.; Vroman, D.; Kim, T.; et al. Current knowledge about and recommendations for ocular methicillin-resistant *Staphylococcus aureus*. *J. Cataract Refract. Surg.* **2014**, *40*, 1894–1908. [CrossRef]
4. Green, M.; Carnt, N.; Apel, A.; Stapleton, F. Queensland microbial keratitis database: 2005–2015. *Br. J. Ophthalmol.* **2019**, *103*, 1481–1486. [CrossRef] [PubMed]
5. Jin, H.; Parker, W.T.; Law, N.W.; Clarke, C.L.; Gisseman, J.D.; Pflugfelder, S.C.; Wang, L.; Al-Mohtaseb, Z.N. Evolving risk factors and antibiotic sensitivity patterns for microbial keratitis at a large county hospital. *Br. J. Ophthalmol.* **2017**, *101*, 1483–1487. [CrossRef]
6. Sand, D.; She, R.; Shulman, I.A.; Chen, D.S.; Schur, M.; Hsu, H.Y. Microbial keratitis in los angeles: The doheny eye institute and the los angeles county hospital experience. *Ophthalmology* **2015**, *122*, 918–924. [CrossRef] [PubMed]
7. Wong, V.W.; Lai, T.Y.; Chi, S.C.; Lam, D.S. Pediatric ocular surface infections: A 5-year review of demographics, clinical features, risk factors, microbiological results, and treatment. *Cornea* **2011**, *30*, 995–1002. [CrossRef]
8. Sweeney, D.F.; Jalbert, I.; Covey, M.; Sankaridurg, P.R.; Vajdic, C.; Holden, B.A.; Sharma, S.; Ramachandran, L.; Willcox, M.D.P.; Rao, G.N. Clinical characterization of corneal infiltrative events observed with soft contact lens wear. *Cornea* **2003**, *22*, 435–442. [CrossRef] [PubMed]
9. Gokhale, N.S. Medical management approach to infectious keratitis. *Indian J. Ophthalmol.* **2008**, *56*, 215–220. [CrossRef]
10. Asbell, P.A.; DeCory, H.H. Antibiotic resistance among bacterial conjunctival pathogens collected in the Antibiotic Resistance Monitoring in Ocular Microorganisms (ARMOR) surveillance study. *PLoS ONE* **2018**, *13*, e0205814. [CrossRef]
11. Morrow, G.L.; Abbott, R.L. Conjunctivitis. *Am. Fam. Physician* **1998**, *57*, 735–746.
12. Monaco, M.; Pimentel de Araujo, F.; Cruciani, M.; Coccia, E.M.; Pantosti, A. *A Worldwide Epidemiology and Antibiotic Resistance of Staphylococcus aureus*; Springer Science and Business Media LLC: Berlin/Heidelberg, Germany, 2016; pp. 21–56.
13. Munita, J.M.; Arias, C.A. Mechanisms of antibiotic resistance. *Microbiol. Spectr.* **2016**, *4*, 4-2. [CrossRef] [PubMed]
14. Ramirez, M.S.; Tolmasky, M.E. Aminoglycoside modifying enzymes. *Drug Resist. Updates* **2010**, *13*, 151–171. [CrossRef] [PubMed]
15. Bush, K. The ABCD's of β-lactamase nomenclature. *J. Infect. Chemother.* **2013**, *19*, 549–559. [CrossRef] [PubMed]
16. Pagès, J.M.; James, C.E.; Winterhalter, M. The porin and the permeating antibiotic: A selective diffusion barrier in Gram-negative bacteria. *Nat. Rev. Microbiol.* **2008**, *6*, 893–903. [CrossRef]
17. Hancock, R.E.; Brinkman, F.S. Function of *Pseudomonas* porins in uptake and efflux. *Annu. Rev. Microbiol.* **2002**, *56*, 17–38. [CrossRef]
18. Poole, K. Efflux-mediated antimicrobial resistance. *J. Antimicrob. Chemother.* **2005**, *56*, 20–51. [CrossRef]
19. Floss, H.G.; Yu, T.W. Rifamycin-mode of action, resistance, and biosynthesis. *Chem. Rev.* **2005**, *105*, 621–632. [CrossRef]
20. Vemula, H.; Ayon, N.J.; Burton, A.; Gutheil, W.G. Antibiotic effects on methicillin-resistant *Staphylococcus aureus* cytoplasmic peptidoglycan intermediate levels and evidence for potential metabolite level regulatory loops. *Antimicrob. Agents Chemother.* **2017**, *61*, e02253-16. [CrossRef]
21. Vemula, H.; Ayon, N.J.; Burton, A.; Gutheil, W.G. Cytoplasmic peptidoglycan intermediate levels in *Staphylococcus aureus*. *Biochimie* **2016**, *121*, 72–78. [CrossRef]
22. Zhang, M.; O'Donoghue, M.M.; Ito, T.; Hiramatsu, K.; Boost, M.V. Prevalence of antiseptic-resistance genes in *Staphylococcus aureus* and coagulase-negative *staphylococci* colonising nurses and the general population in Hong Kong. *J. Hosp. Infect.* **2011**, *78*, 113–117. [CrossRef] [PubMed]
23. Amato, M.; Pershing, S.; Walvick, M.; Tanaka, S. Trends in ophthalmic manifestations of methicillin-resistant *Staphylococcus aureus* (MRSA) in a northern California pediatric population. *J. Am. Assoc. Pediatr. Ophthalmol. Strabis.* **2013**, *17*, 243–247. [CrossRef]

24. Solomon, R.; Donnenfeld, E.D.; Holland, E.J.; Yoo, S.H.; Daya, S.; Güell, J.L.; Mah, F.S.; Scoper, S.V.; Kim, T. Microbial keratitis trends following refractive surgery: Results of the ASCRS infectious keratitis survey and comparisons with prior ASCRS surveys of infectious keratitis following keratorefractive procedures. *J. Cataract Refract. Surg.* 2011, *37*, 1343–1350. [CrossRef]
25. Jensen, S.O.; Lyon, B.R. Genetics of antimicrobial resistance in *Staphylococcus aureus*. *Future Microbiol.* 2009, *4*, 565–582. [CrossRef] [PubMed]
26. Pantosti, A.; Sanchini, A.; Monaco, M. Mechanisms of antibiotic resistance in *Staphylococcus aureus*. *Future Microbiol.* 2007, *2*, 323–334. [CrossRef]
27. Kime, L.; Randall, C.P.; Banda, F.I.; Coll, F.; Wright, J.; Richardson, J.; Empel, J.; Parkhill, J.; O'Neill, A.J. Transient silencing of antibiotic resistance by mutation represents a significant potential source of unanticipated therapeutic failure. *mBio* 2019, *10*, e01755-19. [CrossRef]
28. Vestergaard, M.; Frees, D.; Ingmer, H. Antibiotic resistance and the MRSA problem. *Microbiol. Spectr.* 2019, *7*, 7-2. [CrossRef] [PubMed]
29. Boost, M.; Cho, P.; Wang, Z. Disturbing the balance: Effect of contact lens use on the ocular proteome and microbiome. *Clin. Exp. Optom.* 2017, *100*, 459–472. [CrossRef]
30. Magiorakos, A.P.; Srinivasan, A.; Carey, R.B.; Carmeli, Y.; Falagas, M.E.; Giske, C.G.; Harbarth, S.; Hindler, J.F.; Kahlmeter, G.; Olsson-Liljequist, B.; et al. Multidrug-resistant, extensively drug-resistant and pandrug-resistant bacteria: An international expert proposal for interim standard definitions for acquired resistance. *Clin. Microbiol. Infect.* 2012, *18*, 268–281. [CrossRef] [PubMed]
31. Khan, M.; Stapleton, F.; Willcox, M.D.P. Susceptibility of contact lens-related *Pseudomonas aeruginosa* keratitis isolates to multipurpose disinfecting solutions, disinfectants, and antibiotics. *Transl. Vis. Sci. Technol.* 2020, *9*, 2. [CrossRef]
32. Sauer, A.; Greth, M.; Letsch, J.; Becmeur, P.-H.; Borderie, V.; Daien, V.; Bron, A.; Creuzot-Garcher, C.; Kodjikian, L.; Burillon, C.; et al. Contact lenses and infectious keratitis: From a case-control study to a momputation of the risk for wearers. *Cornea* 2020, *39*, 769–774. [CrossRef] [PubMed]
33. Stapleton, F. Contact lens-related corneal infection in Australia. *Clin. Exp. Optom.* 2020, *103*, 408–417. [CrossRef] [PubMed]
34. MacFadden, D.R.; McGough, S.F.; Fisman, D.; Santillana, M.; Brownstein, J.S. Antibiotic resistance increases with local temperature. *Nat. Clim. Chang.* 2018, *8*, 510–514. [CrossRef] [PubMed]
35. Carmichael, T.R.; Wolpert, M.; Koornhof, H.J. Corneal ulceration at an urban African hospital. *Br. J. Ophthalmol.* 1985, *69*, 920–926. [CrossRef]
36. Upadhyay, M.P.; Karmacharya, P.C.D.; Koirala, S.; Tuladhar, N.R.; Bryan, L.E.; Smolin, G.; Whitcher, J.P. Epidemiologic characteristics, predisposing factors, and etiologic diagnosis of corneal ulceration in Nepal. *Am. J. Ophthalmol.* 1991, *111*, 92–99. [CrossRef]
37. Wahl, J.C.; Katz, H.R.; Abrams, D.A. Infectious keratitis in Baltimore. *Ann. Ophthalmol.* 1991, *23*, 234–237.
38. Goossens, H.; Ferech, M.; Vander Stichele, R.; Elseviers, M. Outpatient antibiotic use in Europe and association with resistance: A cross-national database study. *Lancet* 2005, *365*, 579–587. [CrossRef]
39. Riedel, S.; Beekmann, S.E.; Heilmann, K.P.; Richter, S.S.; Garcia-de-Lomas, J.; Ferech, M.; Goosens, H.; Doern, G.V. Antimicrobial use in Europe and antimicrobial resistance in *Streptococcus pneumoniae*. *Eur. J. Clin. Microbiol. Infect. Dis.* 2007, *26*, 485. [CrossRef]
40. Kantzanou, M. Reduced susceptibility to vancomycin of nosocomial isolates of methicillin-resistant *Staphylococcus aureus*. *J. Antimicrob. Chemother.* 1999, *43*, 729–731. [CrossRef] [PubMed]
41. Chang, V.S.; Dhaliwal, D.K.; Raju, L.; Kowalski, R.P. Antibiotic resistance in the treatment of *Staphylococcus aureus* keratitis: A 20-year review. *Cornea* 2015, *34*, 698–703. [CrossRef]
42. Cabrera-Aguas, M.; Khoo, P.; George, C.R.R.; Lahra, M.M.; Watson, S.L. Antimicrobial resistance trends in bacterial keratitis over 5 years in Sydney, Australia. *Clin. Exp. Ophthalmol.* 2020, *48*, 183–191. [CrossRef]
43. Freidlin, J.; Acharya, N.; Lietman, T.M.; Cevallos, V.; Whitcher, J.P.; Margolis, T.P. Spectrum of eye disease caused by methicillin-resistant *Staphylococcus aureus*. *Am. J. Ophthalmol.* 2007, *144*, 313–315. [CrossRef]
44. Kwiecinski, J.; Jin, T.; Josefsson, E. Surface proteins of *Staphylococcus aureus* play an important role in experimental skin infection. *Apmis* 2014, *122*, 1240–1250. [CrossRef] [PubMed]
45. Watson, S.; Cabrera-Aguas, M.; Khoo, P.; Pratama, R.; Gatus, B.J.; Gulholm, T.; El-Nasser, J.; Lahra, M.M. Keratitis antimicrobial resistance surveillance program, Sydney, Australia: 2016 Annual Report. *Clin. Exp. Ophthalmol.* 2019, *47*, 20–25. [CrossRef] [PubMed]
46. Leibovitch, I.; Lai, T.F.; Senarath, L.; Hsuan, J.; Selva, D. Infectious keratitis in South Australia: Emerging resistance to cephazolin. *Eur. J. Ophthalmol.* 2005, *15*, 23–26. [CrossRef] [PubMed]
47. Samarawickrama, C.; Chan, E.; Daniell, M. Rising fluoroquinolone resistance rates in corneal isolates: Implications for the wider use of antibiotics within the community. *Healthc. Infect.* 2015, *20*, 128–133. [CrossRef]
48. Ly, C.N.; Pham, J.N.; Badenoch, P.R.; Bell, S.M.; Hawkins, G.; Rafferty, D.L.; McClellan, K.A. Bacteria commonly isolated from keratitis specimens retain antibiotic susceptibility to fluoroquinolones and gentamicin plus cephalothin. *Clin. Exp. Ophthalmol.* 2006, *34*, 44–50. [CrossRef]
49. Pratt, R.; Barton, M.; Hart, W. Antibiotic Resistance in Animals. *Commun. Dis. Intell. Q. Rep.* 2003, *27*, S121–S126.
50. Smith, T.C.; Gebreyes, W.A.; Abley, M.J.; Harper, A.L.; Forshey, B.M.; Male, M.J.; Martin, H.W.; Molla, B.Z.; Sreevatsan, S.; Thakur, S.; et al. Methicillin-resistant *Staphylococcus aureus* in pigs and farm workers on conventional and antibiotic-free swine farms in the USA. *PLoS ONE* 2013, *8*, e63704. [CrossRef]

51. Graham, J.P.; Evans, S.L.; Price, L.B.; Silbergeld, E.K. Fate of antimicrobial-resistant *enterococci* and *staphylococci* and resistance determinants in stored poultry litter. *Environ. Res.* **2009**, *109*, 682–689. [CrossRef]
52. Price, L.B.; Johnson, E.; Vailes, R.; Silbergeld, E. Fluoroquinolone-resistant campylobacter isolates from conventional and antibiotic-free chicken products. *Environ. Health Perspect.* **2005**, *113*, 557–560. [CrossRef]
53. Ventola, C.L. The antibiotic resistance crisis: Part 1: Causes and threats. *Pharm. Ther.* **2015**, *40*, 277–283.
54. Marshall, B.M.; Levy, S.B. Food Animals and Antimicrobials: Impacts on Human Health. *Clin. Microbiol. Rev.* **2011**, *24*, 718–733. [CrossRef]
55. Spellberg, B.; Gilbert, D.N. The future of antibiotics and resistance: A tribute to a career of leadership by John Bartlett. *Clin. Infect. Dis.* **2014**, *59* (Suppl. 2), S71–S75. [CrossRef]
56. Sharma, S. Antibiotic resistance in ocular bacterial pathogens. *Indian J. Med. Microbiol.* **2011**, *29*, 218–222. [CrossRef]
57. Dave, S.B.; Toma, H.S.; Kim, S.J. Changes in ocular flora in eyes exposed to ophthalmic antibiotics. *Ophthalmology* **2013**, *120*, 937–941. [CrossRef] [PubMed]
58. Thomas, R.K.; Melton, R.; Asbell, P.A. Antibiotic resistance among ocular pathogens: Current trends from the ARMOR surveillance study (2009–2016). *Clin. Optom.* **2019**, *11*, 15–26. [CrossRef] [PubMed]
59. Marangon, F.B.; Miller, D.; Muallem, M.S.; Romano, A.C.; Alfonso, E.C. Ciprofloxacin and levofloxacin resistance among methicillin-sensitive *staphylococcus aureus* isolates from keratitis and conjunctivitis. *Am. J. Ophthalmol.* **2004**, *137*, 453–458. [CrossRef]
60. Sharma, V.; Sharma, S.; Garg, P.; Rao, G.N. Clinical resistance of *Staphylococcus* keratitis to ciprofloxacin monotherapy. *Indian J. Ophthalmol.* **2004**, *52*, 287–292. [PubMed]
61. Stapleton, F.; Carnt, N. Contact lens-related microbial keratitis: How have epidemiology and genetics helped us with pathogenesis and prophylaxis. *Eye* **2012**, *26*, 185–193. [CrossRef]
62. Chalita, M.R.; Höfling-Lima, A.L.; Paranhos, A.; Schor, P.; Belfort, R. Shifting trends in in vitro antibiotic susceptibilities for common ocular isolates during a period of 15 years. *Am. J. Ophthalmol.* **2004**, *137*, 43–51. [CrossRef]
63. Richards, D.M.; Brogden, R.N. Ceftazidime. *Drugs* **1985**, *29*, 105–161. [CrossRef]
64. Banerjee, R.; Gretes, M.; Harlem, C.; Basuino, L.; Chambers, H.F. A *mecA*-negative strain of methicillin-resistant *Staphylococcus aureus* with high-level β-lactam resistance contains mutations in three genes. *Antimicrob. Agents Chemother.* **2010**, *54*, 4900–4902. [CrossRef]
65. Schubert, T.L.; Hume, E.B.; Willcox, M.D. *Staphylococcus aureus* ocular isolates from symptomatic adverse events: Antibiotic resistance and similarity of bacteria causing adverse events. *Clin. Exp. Optom.* **2008**, *91*, 148–155. [CrossRef]
66. Tuft, S.J.; Matheson, M. In vitro antibiotic resistance in bacterial keratitis in London. *Br. J. Ophthalmol.* **2000**, *84*, 687–691. [CrossRef] [PubMed]
67. Chapman, J. Disinfectant resistance mechanisms, cross-resistance, and co-resistance. *Int. Biodeterior. Biodegrad.* **2003**, *51*, 271–276. [CrossRef]
68. Murray, I.A.; Shaw, W.V. O-Acetyltransferases for chloramphenicol and other natural products. *Antimicrob. Agents Chemother.* **1997**, *41*, 1–6. [CrossRef] [PubMed]
69. Schwarz, S.; Kehrenberg, C.; Doublet, B.; Cloeckaert, A. Molecular basis of bacterial resistance to chloramphenicol and florfenicol. *FEMS Microbiol. Rev.* **2004**, *28*, 519–542. [CrossRef] [PubMed]
70. Wright, G.D. Bacterial resistance to antibiotics: Enzymatic degradation and modification. *Adv. Drug Deliv. Rev.* **2005**, *57*, 1451–1470. [CrossRef] [PubMed]
71. Shaw, W.V. Chloramphenicol acetyltransferase: Enzymology and molecular biology. *Crit. Rev. Biochem.* **1983**, *14*, 1–46. [CrossRef] [PubMed]
72. Wallace, D.C.; Bunn, C.L.; Eisenstadt, J.M. Cytoplasmic transfer of chloramphenicol resistance in human tissue culture cells. *J. Cell Biol.* **1975**, *67*, 174–188. [CrossRef]
73. Chuang, C.-C.; Hsiao, C.-H.; Tan, H.-Y.; Ma, D.H.-K.; Lin, K.-K.; Chang, C.-J.; Huang, Y.-C. *Staphylococcus aureus* ocular infection: Methicillin-resistance, clinical features, and antibiotic susceptibilities. *PLoS ONE* **2012**, *8*, e42437. [CrossRef]
74. Kowalski, R.P.; Kowalski, T.A.; Shanks, R.M.; Romanowski, E.G.; Karenchak, L.M.; Mah, F.S. In vitro comparison of combination and monotherapy for the empiric and optimal coverage of bacterial keratitis based on incidence of infection. *Cornea* **2013**, *32*, 830–834. [CrossRef]
75. Cosgrove, S.E.; Sakoulas, G.; Perencevich, E.N.; Schwaber, M.J.; Karchmer, A.W.; Carmeli, Y. Comparison of mortality associated with methicillin-resistant and methicillin-susceptible *Staphylococcus aureus* bacteremia: A meta-analysis. *Clin. Infect. Dis.* **2003**, *36*, 53–59. [CrossRef]
76. McDonnell, G.; Russell, A.D. Antiseptics, and disinfectants: Activity, action, and resistance. *Clin. Microbiol. Rev.* **1999**, *12*, 147–179. [CrossRef] [PubMed]
77. Clavet, C.R.; Chaput, M.P.; Silverman, M.D.; Striplin, M.; Shoff, M.E.; Lucas, A.D.; Hitchins, V.M.; Eydelman, M.B. Impact of contact lens materials on multipurpose contact lens solution disinfection activity against *Fusarium Solani*. *Eye Contact Lens* **2012**, *38*, 379–384. [CrossRef]
78. Shoff, M.E.; Lucas, A.D.; Brown, J.N.; Hitchins, V.M.; Eydelman, M.B. The effects of contact lens materials on a multipurpose contact lens solution disinfection activity against *Staphylococcus aureus*. *Eye Contact Lens* **2012**, *38*, 368–373. [CrossRef]

79. Codling, C.E.; Maillard, J.Y.; Russell, A.D. Aspects of the antimicrobial mechanisms of action of a polyquaternium and an amidoamine. *J. Antimicrob. Chemother.* **2003**, *51*, 1153–1158. [CrossRef]
80. Ruiz-Linares, M.; Ferrer-Luque, C.M.; Arias-Moliz, T.; de Castro, P.; Aguado, B.; Baca, P. Antimicrobial activity of alexidine, chlorhexidine and cetrimide against *Streptococcus* mutant's biofilm. *Ann. Clin. Microbiol. Antimicrob.* **2014**, *13*, 41. [CrossRef] [PubMed]
81. Abjani, F.; Khan, N.A.; Jung, S.Y.; Siddiqui, R. Status of the effectiveness of contact lens disinfectants in Malaysia against keratitis-causing pathogens. *Exp. Parasitol.* **2017**, *183*, 187–193. [CrossRef] [PubMed]
82. Laxmi Narayana, B.; Rao, P.; Bhat, S.; Vidyalakshmi, K. Comparison of the Antimicrobial Efficacy of Various Contact Lens Solutions to Inhibit the Growth of *Pseudomonas aeruginosa* and *Staphylococcus aureus*. *Int. J. Microbiol.* **2018**, *2018*, 5916712. [CrossRef]
83. Gabriel, M.M.; McAnally, C.; Bartell, J. Antimicrobial efficacy of multipurpose disinfecting solutions in the presence of contact lenses and lens cases. *Eye Contact Lens* **2018**, *44*, 125–131. [CrossRef]
84. Boost, M.V.; Chan, J.; Shi, G.S.; Cho, P. Effect of multipurpose solutions against *Acinetobacter* carrying QAC genes. *Optom. Vis. Sci.* **2014**, *91*, 272–277. [CrossRef] [PubMed]
85. Watanabe, K.; Zhu, H.; Willcox, M. Susceptibility of *Stenotrophomonas maltophilia* clinical isolates to antibiotics and contact lens multipurpose disinfecting solutions. *Investig. Ophthalmol. Vis. Sci.* **2014**, *55*, 8475–8479. [CrossRef] [PubMed]
86. CLSI. *M100 Performance Standards for Antimicrobial Susceptibility Testing*; CLSI: Wayne, PA, USA, 2018.
87. Asbell, P.A.; Sanfilippo, C.M.; Pillar, C.M. Antibiotic resistance among ocular pathogens in the United States. *JAMA Ophthalmol.* **2015**, *133*, 1445. [CrossRef] [PubMed]
88. EUCAST. Clinical Breakpoints and Dosing of Antibiotics. 2018. Available online: https://eucast.org/ast_of_bacteria (accessed on 2 October 2021).
89. Taylor, P.C.; Schoenknecht, F.D.; Sherris, J.C.; Linner, E.C. Determination of minimum bactericidal concentrations of oxacillin for *Staphylococcus aureus*: Influence and significance of technical factors. *Antimicrob. Agents Chemother.* **1983**, *23*, 142–150. [CrossRef]

Article

Genomics of *Staphylococcus aureus* Strains Isolated from Infectious and Non-Infectious Ocular Conditions

Madeeha Afzal *, Ajay Kumar Vijay, Fiona Stapleton * and Mark D. P. Willcox *

School of Optometry and Vision Science, University of New South Wales, Sydney, NSW 2052, Australia; v.ajaykumar@unsw.edu.au
* Correspondence: m.afzal@unsw.edu.au (M.A.); f.stapleton@unsw.edu.au (F.S.); m.willcox@unsw.edu.au (M.D.P.W.)

Abstract: *Staphylococcus aureus* is a major cause of ocular infectious (corneal infection or microbial keratitis (MK) and conjunctivitis) and non-infectious corneal infiltrative events (niCIE). Despite the significant morbidity associated with these conditions, there is very little data about specific virulence factors associated with the pathogenicity of ocular isolates. A set of 25 *S. aureus* infectious and niCIEs strains isolated from USA and Australia were selected for whole genome sequencing. Sequence types and clonal complexes of *S. aureus* strains were identified by using multi-locus sequence type (MLST). The presence or absence of 128 virulence genes was determined by using the virulence finder database (VFDB). Differences between infectious (MK + conjunctivitis) and niCIE isolates from USA and Australia for possession of virulence genes were assessed using the chi-square test. The most common sequence types found among ocular isolates were ST5, ST8 while the clonal complexes were CC30 and CC1. Virulence genes involved in adhesion (*ebh*, *clfA*, *clfB*, *cna*, *sdrD*, *sdrE*), immune evasion (*chp*, *esaD*, *esaE*, *esxB*, *esxC*, *esxD*), and serine protease enzymes (*splA*, *splD*, *splE*, *splF*) were more commonly observed in infectious strains (MK + conjunctivitis) than niCIE strains ($p = 0.004$). Toxin genes were present in half of infectious (49%, 25/51) and niCIE (51%, 26/51) strains. USA infectious isolates were significantly more likely to possess *splC*, *yent1*, *set9*, *set11*, *set36*, *set38*, *set40*, *lukF-PV*, and *lukS-PV* ($p < 0.05$) than Australian infectious isolates. MK USA strains were more likely to possesses *yent1*, *set9*, *set11* than USA conjunctivitis strains ($p = 0.04$). Conversely USA conjunctivitis strains were more likely to possess *set36 set38*, *set40*, *lukF-PV*, *lukS-PV* ($p = 0.03$) than MK USA strains. The ocular strain set was then compared to 10 fully sequenced non-ocular *S. aureus* strains to identify differences between ocular and non-ocular isolates. Ocular isolates were significantly more likely to possess *cna* ($p = 0.03$), *icaR* ($p = 0.01$), *sea* ($p = 0.001$), *set16* ($p = 0.01$), and *set19* ($p = 0.03$). In contrast non-ocular isolates were more likely to possess *icaD* ($p = 0.007$), *lukF-PV*, *lukS-PV* ($p = 0.01$), *selq* ($p = 0.01$), *set30* ($p = 0.01$), *set32* ($p = 0.02$), and *set36* ($p = 0.02$). The clones ST5, ST8, CC30, and CC1 among ocular isolates generally reflect circulating non-ocular pathogenic *S. aureus* strains. The higher rates of genes in infectious and ocular isolates suggest a potential role of these virulence factors in ocular diseases.

Keywords: *Staphylococcus aureus*; ocular infectious isolates; whole genome sequencing; virulence factors

Citation: Afzal, M.; Vijay, A.K.; Stapleton, F.; Willcox, M.D.P. Genomics of *Staphylococcus aureus* Strains Isolated from Infectious and Non-Infectious Ocular Conditions. *Antibiotics* 2022, 11, 1011. https://doi.org/10.3390/antibiotics11081011

Academic Editor: Hiroshi Hamamoto

Received: 5 July 2022
Accepted: 26 July 2022
Published: 27 July 2022

Publisher's Note: MDPI stays neutral with regard to jurisdictional claims in published maps and institutional affiliations.

Copyright: © 2022 by the authors. Licensee MDPI, Basel, Switzerland. This article is an open access article distributed under the terms and conditions of the Creative Commons Attribution (CC BY) license (https://creativecommons.org/licenses/by/4.0/).

1. Introduction

Staphylococcus aureus is responsible for nearly 70% of ocular infections worldwide [1]. These can result in tissue damage, morbidity, and vision loss [2,3]. *S. aureus* infections involving the cornea (microbial keratitis; MK) can be sight-threatening and the organism is the most common cause of MK in Australia [4,5] and USA [6,7]. *S. aureus* can also cause conjunctivitis [8] and non-infectious corneal infiltrative events (niCIE) during contact lens wear [9].

S. aureus is known to encode a diverse arsenal of virulence determinants that enables it to cause a variety of infections [10]. The genomic make-up of *S. aureus* influences the virulence of its strains and pathogenicity associated with its disease [11]. The antibiotic susceptibility data of the isolates previously reported [12] and used in this study demonstrated that although most of the strains were multi-drug resistant (MDR), the non-infectious (niCIE) strains were more susceptible to antibiotics (ciprofloxacin, ceftazidime, oxacillin) than were the conjunctivitis strains, and the conjunctivitis strains were more susceptible to antibiotics (chloramphenicol, azithromycin) than were the MK strains [12]. MK strains from Australia were more susceptible to antibiotics (ciprofloxacin, oxacillin) compared to MK strains from USA [12]. Whilst several studies have examined which virulence factors might be involved in the development of keratitis by *S. aureus*, there is much less information on the association of virulence factors with conjunctivitis or niCIE [13]. Similarly, as outlined previously [13], infectious isolates (MK + conjunctivitis) had a higher frequency of genes involved in evasion of the immune system and invasion of the host (*hlg*, *hld*) compared to niCIE strains. On the other hand, *scpA*, that encodes a staphylococcal cysteine proteinase, was more common in niCIE strains. However, those previous studies only examined a subset of genes, specifically those that had been previously reported to be involved in infections of the eye or antibiotic resistance. This current study examines the whole genome of a subset of strains isolated from MK, conjunctivitis, and niCIE. This analysis may identify new genes that are associated with particular infections or resistance to antibiotics.

Whole genome sequencing (WGS) is a widely used technique that can identify antibiotic resistance genes, virulence determinants, emerging bacterial lineages, and their population structures [14–16]. Comparative genomics and genome-wide association studies of clinical isolates can reveal genetic determinants that may be important in the setting of specific infections. For example, WGS has been used successfully to examine *S. aureus* isolates collected from systemic infections (bloodstream, airways, endocarditis, and joint infections) to further understand specific population structures as well as to explore the relationship between virulence factors and patient outcomes [16–19].

WGS of *S. aureus* strains isolated from different infections (airways, soft-tissues, and skin lesions) showed high level of diversity and co-presence of local, global, livestock-associated, and hypervirulent clones and found that some virulence factors and clones were disease specific. For example, the sequence type ST22 was associated with toxic shock syndrome toxin TSST-1 and ST5 was associated with enterotoxins (SE) [18]. Another study explored genomic relatedness between commensal nasal isolates and those isolated from prosthetic joint infections and found the commensals shared the same clonal complex (CC) and the prevalence of virulence genes among isolates from commensal and prosthetic joint infections in arthroplasty patients was almost equal, suggesting that commensal *S. aureus* nasal clones can cause joint infections [19].

In the current study, WGS was used to analyze 25 *S. aureus* strains from ocular infectious and niCIEs isolated from USA and Australia. A custom analytical pipeline determined MLST, to define circulating *S. aureus* ocular lineages in infectious and non-infectious strains from USA and Australia, as well as the presence or absence of 128 known *S. aureus* virulence factors. The ocular strains were then compared to 10 fully sequenced non-ocular strains to determine the key virulence factors involved in ocular diseases.

2. Results

2.1. General Features of the Genomes

After de novo assembly, the isolates had different numbers of contigs ranging from 328 for SA31 to 3916 for SA86. Isolates had an average guanine plus cytosine (GC) content of 32.8%. The tRNA copy number for the isolates ranged from 60 to 89. Similarly, the number of coding sequences (CDS), which was determined based on Prokka annotation pipeline, ranged from 2614 (in M19-01) to 3873 (SA86). The general features of isolates are provided in Table 1.

Table 1. Genetic features of the S. aureus isolates.

Ocular Condition	S. aureus Isolates	Region	GC Content (%)	No. of Contigs	Total Sequence Length (bp)	CDSs (Total)	tRNAs
Microbial keratitis	SA107	USA	32.8	1173	3,599,003	3302	71
	SA111		33	655	3,113,006	2858	85
	SA112		32.9	614	3,170,760	2930	74
	SA113		32.8	530	3,014,859	2771	72
	SA114		32.9	1332	3,175,242	2877	60
	SA34	AUS	32.9	349	2,914,342	2694	60
	SA129		32.9	694	3,105,791	2897	66
	M5-01		32.9	624	2,975,620	2701	85
	M19-01		33	429	2,893,905	2614	77
	M28-01		32.8	475	2,960,866	2715	62
	M43-01		33.1	985	3,029,867	2741	89
	M71-01		32.9	536	2,918,758	2665	74
Conjunctivitis	SA86	USA	32.9	3916	4,579,417	3873	76
	SA90		32.8	404	3,015,554	2755	62
	SA101		32.6	998	3,602,977	3296	63
	SA102		32.8	1067	3,406,253	3085	65
	SA103		32.9	479	3,069,147	2857	72
	SA46	AUS	32.9	388	2,903,724	2646	62
	SA136		32.8	735	3,035,909	2803	76
niCIE	SA20	AUS	32.8	385	2,909,603	2660	61
	SA25		32.8	366	2,907,754	2622	61
	SA27		32.8	345	2,919,830	2686	67
	SA31		32.8	328	2,976,006	2782	60
	SA32		32.7	649	2,990,036	2665	65
	SA48		32.8	338	2,922,947	2665	64

CDS = coding DNA sequence. Note: all strains had N50 values of 985.

2.2. Acquired Antimicrobial Resistance Genes

Eighteen different types of acquired antimicrobial resistance genes for various classes of antibiotics were detected in this study (Table 2). Antimicrobial resistance genes for vancomycin (*vanA*), fusidic acid (*fusA*, *fusB*), trimethoprim (*dfrA*, *dfrB*, *dfrG*), ciprofloxacin (*gyrA*, *gyrlA*, *grlB*), fosfomycin (*fosB*), and rifampin (*rpoB*) were not found in any of the strains. The beta lactamase resistance gene *blaZ* which encodes penicillin resistance was found in 76% of isolates. However, the methicillin resistance gene *mecA* was found in 28% of strains, all of which were from the USA; the possession of *mecA* was significantly more common in infectious isolates from USA than from Australia ($p = 0.0016$).

The aminoglycoside resistance genes were significantly more common ($p = 0.0006$) in strains from the USA, with only strain M28-01 isolated from MK in Australia possessing one of these genes, *ant(9)-la*. Genes associated with resistance to macrolides, lincosamide, or streptogramin B were significantly more likely to be found in USA isolates ($p = 0.002$), with only Australian isolates M28-01 possessing *erm(A)* and SA25 possessing *erm(A)*, *msr(A)*, and *erm(C)*. Six isolates possessed *tetK* that encodes tetracycline resistance, and these were scattered across isolates from MK (2 USA, 1 Australian), conjunctivitis (1 USA) and niCIE (2 Australia). Resistance gene for tetracycline (*tetM*) and quaternary ammonium compound (*qacD*) were found in single isolate (USA) whereas pseudomonic acid (mupirocin) was present in only two USA isolates and quaternary ammonium compound *qacB* was found in single USA and single Australian isolate. Chloramphenicol resistance gene *cat(pC233)* was only found in a single Australian isolate

Overall, in Australian infectious isolates only five acquired antimicrobial resistance genes were detected. As the current study relied on draft genomes it may not be able to predict actual genomic diversity and could not detect actual antimicrobial resistance genes. There could be more genes, complete gene sequence of isolates can show the actual number of antimicrobial resistance genes. Similarly, USA infectious isolates had acquired 17 different antimicrobial resistance genes (Table 2). NiCIE isolates from Australia had acquired six different antimicrobial resistance genes. One USA infectious isolate, SA101, had the largest number of acquired antimicrobial resistance genes (eight). (Table 2).

Table 2. Acquired antimicrobial resistance genes in S. aureus isolates from different ocular conditions.

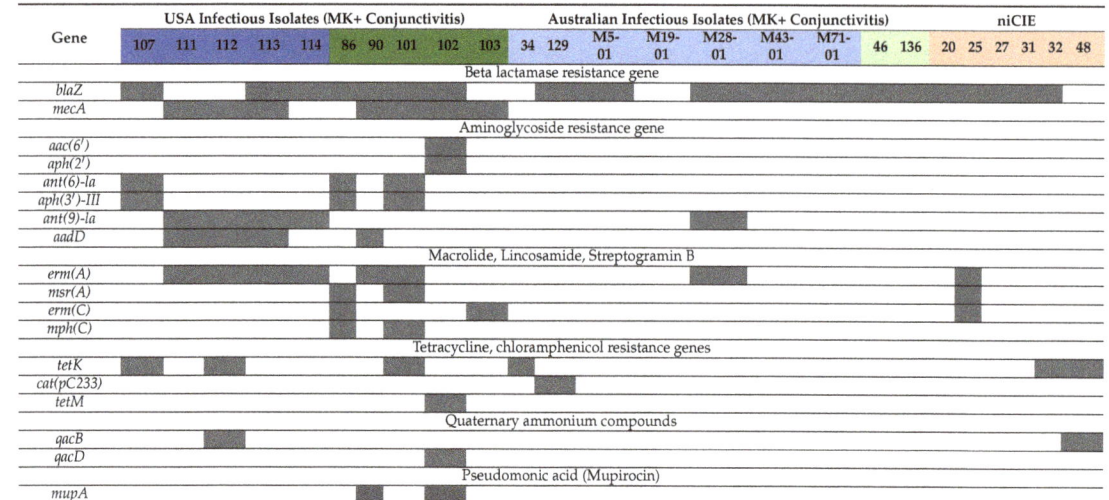

Grey color represents the presence of the gene. Dark blue = USA MK strains, light blue = Australian MK strains; dark green = conjunctivitis USA strains, light green = conjunctivitis Australian strain, peach color = niCIE strains.

2.3. S. aureus Virulence Determinants

Of the 128 virulence factors examined, 22 virulence genes (*atl, ebh, clfA, clfB, cna, ebp, eap, efb, fnbA, fnbB, icaA, icaB, icaC, icaD, icaR, sdrC, sdrD, sdrE, sdrF, sdrG, sdrH, spa*) in VFDB are categorized as genes involved in *S. aureus* adhesion. Of these adhesins, *atl, ebp, eap, efb, fnbA, fnbB, icaA, icaB, icaC, icaR, sdrC,* and *spa* were found in ≥96% of all *S. aureus* isolates. On the other hand, *sdrF, sdrG, sdrH* were not detected in any of the strains.

S. aureus strains from ocular infectious and niCIE showed non-significant differences in the frequency of possession of six adhesins (Figure 1), with only possession of *icaD* showing a trend towards being more common in niCIE isolates ($p = 0.1$).

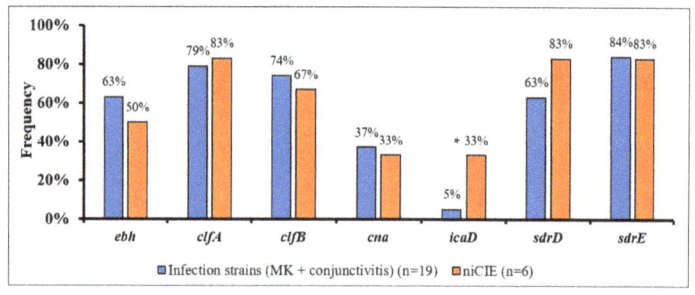

Figure 1. Frequency of seven virulence genes involved in *S. aureus* adhesion by disease group. *, trend to be more common in niCIE strains ($p = 0.1$).

When differences were examined for the possession of adhesins in the infectious isolates from different countries, there were no significant differences observed in MK and conjunctivitis isolates from USA and Australia.

Of the remaining 106 virulence genes, 15 were categorized as enzymes in VFDB. These were genes for the cysteine proteases *scpA* and *sspB*, the serine proteases *sspA, splA, splB, splC, slpD, splE, splF*, hyaluronate lyase *hysA*, the lipases *geh* and *lip*, staphylocoagulase *coa*, staphylokinase *sak* and thermonuclease *nuc*. Of these genes, 67% (9/15; *sspB, hysA, geh, lip, v8, sspA, sak,* and *nuc*) were found in ≥96% of all *S. aureus* isolates. The isolates from infections or niCIEs isolates did not show significant differences ($p > 0.05$) or trend towards significance ($p = 0.1$), for the possession of other seven proteases, (Figure 2).

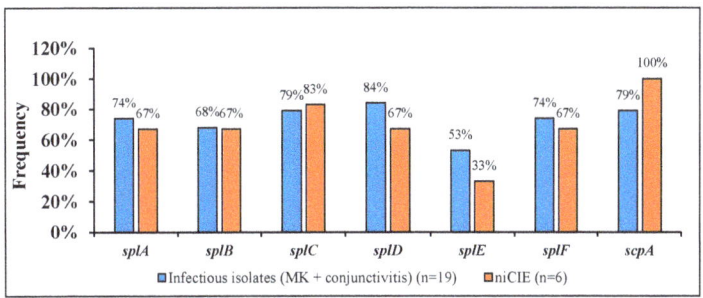

Figure 2. Frequency of 7 proteases in *S. aureus* by disease group.

Similarly, the frequency of protease genes in strains isolated from MK or conjunctivitis was not significantly different, although conjunctivitis strains (100%) had a trend for more frequent presence of *splA* ($p = 0.1$) and *splF* ($p = 0.1$) than MK strains (58%). In infectious isolates (MK + conjunctivitis) from the USA, possession of *splC* (100% vs. 55%; $p = 0.03$) and *splB* (90% vs. 44%; $p = 0.05$) was higher and there was also a trend for higher possession of *splD* (100% vs. 66%; $p = 0.08$) and *splA* (100% vs. 55%; $p = 0.1$) compared to Australian isolates, except *scpA* (100% vs. 60%; $p = 0.08$) which was higher in infectious isolates (MK + conjunctivitis) from Australia.

Of the remaining 91 virulence genes, five were involved in immune evasion (IE; *adsA, chp, cpsA, scn, sbi*), and 12 genes were involved in the type VII secretion systems (*esaA, esaB, esaD, esaE, esaG, essA, essB, essC, esxA, esxB, esxC, esxD*). Of these, 10/17 (*adsA, cpsA, scn, sbi, esaA, esaG, essA, essB, essC,* and *esxA*) were found in ≥96% of all *S. aureus* isolates. There were no significant differences or trends in possession of any IE or type VII secretion system genes by disease group or by country. Figure 3 shows the differences in possession of seven of these genes between infectious and niCIE isolates.

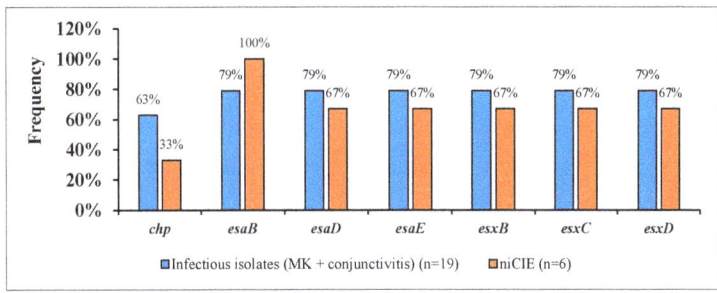

Figure 3. Possession of 7 virulence genes involved in immune evasion and type VII secretion system in *S. aureus* by disease group.

The remaining 74 virulence genes encoded for toxins including hemolysins ($hla, hlb, hld, hlgA, hlgB, hlgC$), enterotoxins (*sea, seb, sec, sed, see, seg, seh, sei, sej, yent1, yent2, selk, sell, selm, seln, selo, selp, selq, selr, selu*), exfoliative toxins (*eta, etb, etc, etd*), exotoxins, also known as enterotoxin like genes, (*set1, set2, set3, set4, set5, set6, set7, set8, set9, set10, set11, set12, set13, set14, set15, set16, set17, set18, set19, set20, set21, set22, set23, set24, set25, set26, set30, set31, set32, set33, set34, set35, set36, set37, set38, set39, set40*), leukocidins (*lukF-like, lukM, lukD, lukE, lukf-PV, lukS-PV*), and toxic shock syndrome toxin (*tsst*).

Of these, the hemolysins *hla, hlgA, hlgB, hlgC* were found in ≥96% of all *S. aureus* isolates. Of the remaining 70 toxins, *sed, see, sej, selp, selr, eta, etb, etc, etd, set10, set12, set14, set20, lukM* were not detected in any of the isolates, and *hlb, sell, set35, lukF-like, lukE* were present only in 4% of all *S. aureus* isolates. However, 51 toxins showed some differences between *S. aureus* infectious and niCIE isolates (Figure 4). Of these the only significant differences or trends for differences were as follows: niCIE isolates tended to have a higher frequency (50%) of only *set3* ($p = 0.1$) (Figure 4) than infectious isolates (16%), and infectious isolates tended to have a higher frequency (95%) of only *hld* ($p = 0.1$) than niCIE (67%) (Figure 5).

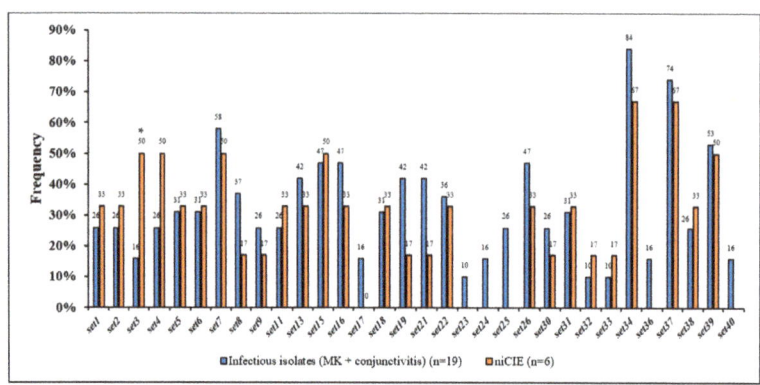

Figure 4. Frequency of 32 enterotoxin-like genes in *S. aureus* by disease group. *, trend more common in niCIE strains ($p = 0.1$).

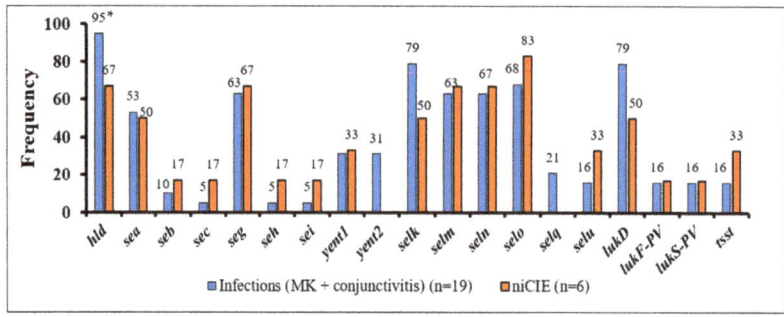

Figure 5. Frequency of 19 enterotoxins, exfoliative toxins and *tsst* in *S. aureus* by disease group. *, trend more common in infectious strains ($p = 0.1$).

Overall conjunctivitis strains were more likely to possess *set36* (43% vs. 0%; $p = 0.03$), *set38* (57% vs. 8%; $p = 0.03$), *set40* (43% vs. 0%; $p = 0.03$), *lukF-PV* (43% vs. 0%; $p = 0.03$), *lukS-PV* (43% vs. 0%; $p = 0.03$), with a trend for *set31* (57% vs. 17%; $p = 0.1$) than MK strains. These 51 toxins were also examined for differences in the isolate's country of origin. The only differences were for MK strains, where isolates from the USA had a significantly higher frequency of possession of *yent1* (60% vs. 0%; $p = 0.04$), *set9* (60% vs. 0%; $p = 0.04$), and *set11* (60% vs. 0%; $p = 0.04$) than MK isolates from Australia.

The VFDB results of these 25 ocular isolates were compared with previously published non-ocular isolates for the possession of the 128 virulence determinants. Eight genes involved in adhesion (*ebp, eap, efb, fnbA, fnbB, icaA, icaR, sdrC*) were found in all ocular and non-ocular *S. aureus* isolates and three (*sdrF, sdrG, sdrH*) were not found in any isolate. *S. aureus* ocular isolates showed higher frequency for the possession of *cna* (40% vs. 0%; $p = 0.03$) and *icaR* (100% vs. 70%; $p = 0.01$) whereas non-ocular isolates showed higher frequency for the possession of *icaD* (60% vs. 12%; *p* = 0.007), *ebh* (90% vs. 60%; *p* = 0.1), and *sdrD* (100% vs. 68%; *p* = 0.07). Of 15 enzymes, 9 (*spa, sspB, sspC, hysA, geh, lip, sspA, coa, nuc*) were found in all *S. aureus* isolates and no significant differences (or trends) were found in the possession of any other enzyme-associated gene. Similarly, all five (*adsA, cpsA, scn, sbi, chp*) genes involved in immune evasion were found in all isolates. Six (*esaA, esaB, esaG, essA, essB, esxA*) of the genes involved in type VII secretion system were found in ≥96% of all *S. aureus* isolates, with non-significant differences in frequency of possession of the remaining six, type VII secretion system gene, *esaD* (80% vs. 100%; *p* = 0.29), *esaE* (80% vs. 100%; *p* = 0.29), *essC* (92% vs. 100%; *p* = 0.99), *esxB* (76% vs. 100%; *p* = 0.15), *esxC* (76% vs. 100%; *p* = 0.15), *esxD* (76% vs. 100%; *p* = 0.15) in ocular and non-ocular isolates respectively.

Of 74 toxins, four hemolysin genes (*hla, hlgA, hlgB, hlgC*) were found in ≥96% of all *S. aureus* isolates and 18 toxin genes (*hlb, sed, see, sej, selp, selr, eta, etb, etc, etd, set10, set12, set14, set20, set35, lukF-like, lukM, lukE*) were found in ≤4% of all strains. Of the remaining 52 toxins, *S. aureus* ocular isolates possessed *sea* (80% vs. 20%; *p* = 0.001), *set1* (28% vs. 0%; *p* = 0.08), *set5* (32% vs. 0%; *p* = 0.07), *set16* (44% vs. 0%; *p* = 0.01), *set19* (36% vs. 0%; *p* = 0.03), whereas non-ocular isolates possessed *selq* (30% vs. 0%; *p* = 0.018), *set30* (70% vs. 24%; *p* = 0.01), *set32* (50% vs. 12%; *p* = 0.02), *set36* (50% vs. 12%; *p* = 0.02), *set37* (100% vs. 72%; *p* < 0.0001), *lukD* (100% vs. 72%; *p* < 0.0001), *lukF-PV* (60% vs. 16%; *p* = 0.001), and *lukS-PV* (60% vs. 16%; *p* = 0.001).

2.4. Sequence Types and Clonal Complexes of S. aureus Isolates

The MLST typing of 25 *S. aureus* genomes revealed a total of 14 distinct sequence types (STs) and seven clonal complexes (Table 3), ST5 (n = 5, 20%) and ST8 (n = 4, 16%) were the most common sequence types in this cohort of ocular isolates. For strain M19-01, no ST type was identified, and was named as NI (Table 3). In the current study most of the USA isolates were from CC5 or CC8, whereas there was a greater spread of sequence types and clonal complexes in the Australian isolates.

The core, shell (genes present in two or more strains), and pan genes of published isolates are provided in Supplementary Table S1. The core genes were used to create a phylogenetic tree of the *S. aureus* isolates using *S. aureus* NCTC 8325 (NC_007795.1) as a reference strain. The ten published non-ocular *S. aureus* isolates downloaded from the Genebank database were also included. Isolates of the same clonal complex or same sequence type were grouped together in the same cluster irrespective of their ocular condition or country of origin (Figure 6). The core genomes formed three groups in the phylogenetic tree (Figure 6). Isolates in Group 1 were related, as they belonged to the same CC5. This group also contained all the extensively-drug resistant isolates (XDR: resistant to almost all antibiotics) (SA111, SA112, SA113) and three multi-drug resistant (MDR: resistant to three different classes of antibiotics) isolates (SA90, SA48, SA46) reported in the previous study [12]. Isolates from CC30 in Group 2 were further clustered into two sub-lineages based on their core genes and STs. Group 3 was larger and contained the majority of MDR strains. Australian strains clustered into sub lineages within group 3, whereas USA isolates with same ST8 clustered together within group 3.

Table 3. Sequence types and clonal complexes of *S. aureus* isolates.

S. aureus Isolates	Sequence Type	Number of: Clonal Complex	Core Genes	Shell Genes	Pan/Total Genes
107	15	CC15	2392	1187	3579
111	105	CC5	2382	770	3152
112	5	CC5	2380	841	3221
113	105	CC5	2330	782	3112
114	30	CC30	2168	129	3377
86	840	CC5	1984	2577	4561
90	5	CC5	2342	739	3089
101	8	CC8	2533	898	3431

Table 3. *Cont.*

S. aureus Isolates	Sequence Type	Number of: Clonal Complex	Core Genes	Shell Genes	Pan/Total Genes
102	8	CC8	2497	760	3257
103	8	CC8	2514	498	3012
34	508	CC45	2194	974	3168
129	34	CC30	2227	1112	3339
M5-01	188	CC1	2267	844	3111
M19-01	NI	NI	2302	684	2986
M28-01	109	CC1	2304	775	3079
M43-01	672	NI	2296	827	3123
M71-01	97	CC97	2315	705	3020
46	5	CC5	2325	664	2989
136	188	CC1	2328	821	3149
20	121	NI	2252	824	3076
25	5	CC5	2341	608	2949
27	39	CC30	2180	996	3176
31	34	CC30	2220	1010	3230
32	8	CC8	2416	501	2917
48	5	CC5	2300	736	3036

NI = not identified. Isolates highlighted in shades of blue indicate MK strains; dark blue represents MK strains from USA and light blue represents MK strains from Australia. Shades of green indicate conjunctivitis strains; dark green indicates conjunctivitis strains from USA and light green indicates conjunctivitis strains from Australia. The peach color indicates strains from niCIE.

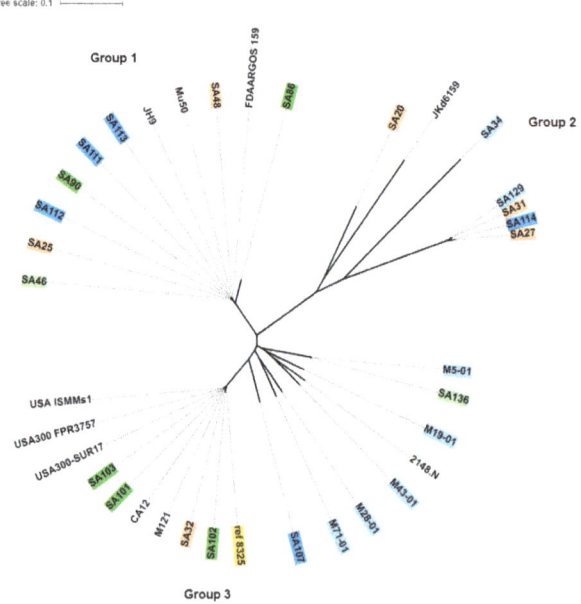

Figure 6. Core genome phylogeny of *S. aureus*, using Parsnp. *S. aureus* strain NCTC 8325, was used as a reference strain (yellow). Isolates highlighted in shades of green indicate conjunctivitis strains; dark green indicates conjunctivitis strains from USA and light green indicates conjunctivitis strains from Australia. Shades of blue indicate MK strains; dark blue represents MK strains from USA and light blue represents MK strains from Australia. The peach color indicates strains from niCIE and strains with no color indicate non-ocular isolates. The tree was constructed using online webtool itol (interactive tree of life, https://itol.embl.de/ (accessed on 4 April 2022).

The pan phylogenetic relationships of these *S. aureus* isolates were assessed (Figure 7). This divided the *S. aureus* isolates into three major groups. Group 1 of the pan genome phylogeny contained only two isolates, ocular isolate M43-01 and a non-ocular isolate of the same clonal complex. The second group included isolates with the same clonal complex, and XDR (resistant to almost all antibiotics) and MDR (resistant to three different classes of antibiotics) strains, irrespective of their ocular condition and country of origin. Group 3 was further divided into two subgroups; isolates with the same pangenome and CC or ST were clustered together. Isolates in group 3 had a large number of pan genes. Isolates belonging to the same sequence type or clonal complex were grouped together, for example, isolates 129, 31, 114, 27 were from clonal complex 30.

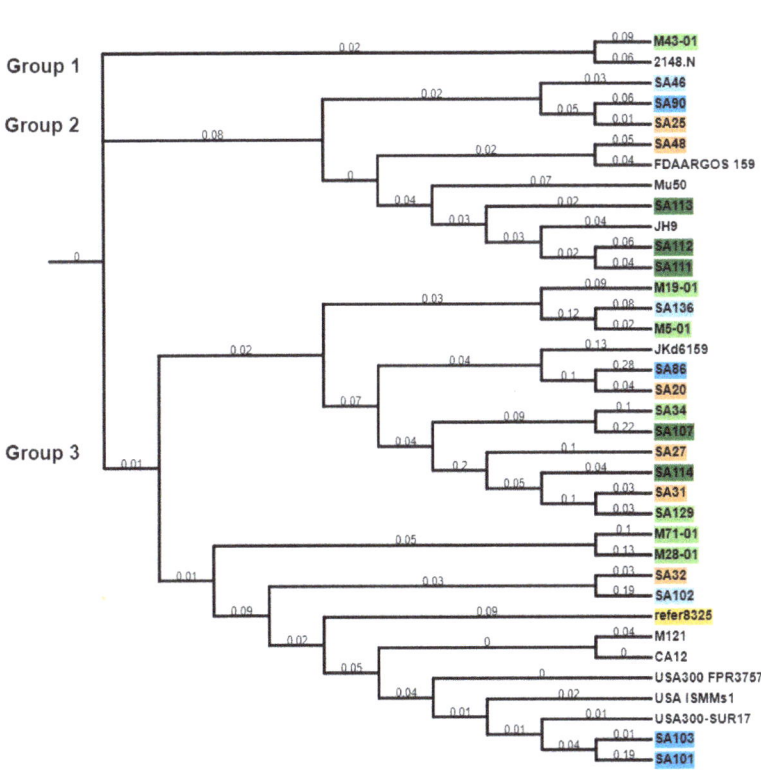

Figure 7. Pan genome phylogeny of *S. aureus* isolates. *S. aureus* strain NCTC 8325 was used as a reference strain (yellow color). Isolates in shades of green are conjunctivitis strains; dark green indicates conjunctivitis strains from USA, and light green conjunctivitis strains from Australia. Shades of blue indicate MK strains; dark blue represents MK strains from USA and light blue represents MK strains from Australia. Isolates in peach color are strains from niCIE and those with no color are non-ocular isolates. The tree was constructed using online webtool itol (interactive tree of life, https://itol.embl.de/, (accessed on 4 April 2022). The tree scale indicates differences between the isolates and branch length indicates the number of changes that have occurred in that branch.

3. Discussion

This study investigated genomic differences in resistance and virulence genes of *S. aureus* isolates from different infectious (MK and conjunctivitis) and non-infectious (niCIE) ocular conditions from USA and Australia. Based on previous phenotypic susceptibility [12] and PCR data [13] it was expected that there would be differences in the resistance and virulence determinants between infectious and non-infectious disease. Most (n = 22, 88%) of the isolates used in the study were MDR [12]. Phenotypically non-infectious (niCIE) isolates in a previous study were more susceptible to antibiotics (ciprofloxacin, ceftazidime and oxacillin) than conjunctivitis and MK strains, and MK isolates from USA were more resistant to antibiotics than MK isolates from Australia (ciprofloxacin, ceftazidime and oxacillin) [12]. The current study's genotypic data shows that infectious isolates from USA harbored more antimicrobial resistance genes (ARGs) compared to Australian isolates, which supports the phenotypic data of the previous study. Similarly, PCR results for a subset of 12 known virulence genes had previously reported that genes involved in evasion and invasion (*hlg* and *hld*) were more commonly found in infectious isolates than niCIE [13]. The current study results for *hld* are consistent with the previous report [13]. However, *hlg* which was more common in infectious strains than niCIE strains in the previous study [13] was found in ≥96% of all *S. aureus* isolates in the current study. In addition, due to the selection of isolates in the current study, staphylococcal cysteine proteinase *scpA*, which was more common in niCIE isolates than infectious strains [13] in the previous study, whilst being more commonly observed in niCIE strains than infectious strains in the current study, did not reach significance (100% vs. 79%, $p = 0.28$).

Overall, 76% of all strains possessed the acquired penicillin resistance gene *blaZ* but only 28% of strains, all from USA infectious (MK+ conjunctivitis), possessed *mecA* (i.e., were MRSA). The high level of MRSA among *S. aureus* ocular isolates from USA in the current study is consistent with previous studies [20,21]. The current study reports low level of MRSA among ocular isolates from Australia which supports previous studies showing low rates (≤6.3%) of MRSA among *S. aureus* ocular isolates from Australia [12,22].

The aminoglycoside resistance genes *aac (6')* and *aph (2')*, which encode for gentamicin resistance, were found in only one isolate from USA which is consistent with phenotypic susceptibility [12] and other previous studies from USA and Australia [22–24] which suggest gentamicin remains a good option to treat *S. aureus* ocular infections in both Australia and USA. Genes *ant (6)-la*, *aph (3') III*, which encode for streptomycin resistance, were found in three USA isolates but in none of the Australian isolates. Streptomycin is no longer used in clinical treatment [25], so this resistance may not be clinically relevant but does suggest environmental selection for the persistence of genes. Gene *ant (9)*-la, which confers resistance to spectinomycin, was found in four USA isolates (three were MRSA) and one isolate from Australia. Several previous reports showed an association between aminoglycoside resistance and methicillin resistance [26,27]. Gene *aadD*, which is responsible for resistance to kanamycin/neomycin and tobramycin [28], was found in four USA isolates.

Strains from niCIE showed a trend of higher frequency possession of *icaD*, the intercellular adhesion gene, is involved in biofilm production [29]. As niCIE are associated with contact lens wear and contact lenses may provide a surface where bacteria can attach and colonize as a biofilm [30], it is perhaps not surprising that possession of *icaD* was more common in niCIE isolates and suggests that biofilm formation mediated by this gene is not critical for ocular surface infection (i.e., MK or conjunctivitis). In the current study when ocular isolates were compared to non-ocular isolates, they showed higher frequency for the possession of *cna* (40% vs. 0%; $p = 0.03$) and *icaR* (100% vs. 70%; $p = 0.01$), whereas non-ocular isolates showed higher frequency for the possession of *icaD* (60% vs. 12%; $p = 0.007$), *ebh* (90% vs. 60%; $p = 0.1$), and *sdrD* (100% vs. 68%; $p = 0.07$). The product of *cna*, collagen binding adhesin, has been reported to be involved in the pathogenesis of *S. aureus* keratitis [31] and the possession of this gene in ocular strains in the current study confirms that it may be an important virulence determinant in *S. aureus* ocular infections. Gene *icaR* is a strong negative regulator of biofilm formation, and its absence enhances PNAG (poly-N-acetylglucosamine) production and biofilm formation [32,33]. Ocular strains used in this study are enriched with *icaR* which further suggests biofilm formation is not an absolute requirement for *S. aureus* ocular infections.

Non-ocular strains in the current study were enriched with *icaD* which is involved in biofilm production [29]; this suggests that biofilm formation is important for their non-ocular pathogenesis. Gene *ebh* is a cell wall-associated fibronectin binding protein [34] which helps *S. aureus* to adhere to host extracellular matrix (ECM) and plays a role in cell growth, envelope assembly [35] while contributing to structural homeostasis of bacterium by forming a bridge between the cell wall and cytoplasmic membrane [36]. The lower frequency of *ebh* possession in *S. aureus* ocular strains suggests it has a minor role in eye infections. Gene *ebh* is produced during human blood infection, as serum samples taken from patients with confirmed *S. aureus* infection were found to contain anti-*ebh* antibodies. Gene *sdrD* (serine–aspartate repeat protein D) is member of the MSCRAMMs (microbial surface components recognizing adhesive matrix molecules) [37], promotes the adherence of *S. aureus* to nasal epithelial cells [38], human keratinocytes [39], and contributes to abscess formation [40]. A high prevalence of *S. aureus sdrD* gene is reported among patients with bone infections [41] which suggests that *sdrD* may contribute to systemic infection. *sdrD* is also reported to aid the pathogen in immune evasion by increasing *S. aureus* virulence and survival in blood [40]. The lower frequency of possession of this gene in ocular isolates indicates *sdrD* may not be involved in pathogenesis of currently prevalent types of eye infections, however, *S. aureus* with *sdrD* could contribute more to eye infection.

Overall, isolates from conjunctivitis had a higher frequency for possession of the serine proteases *splA* and *splF* than isolates from MK. Infectious isolates from USA were significantly, or trended to be more likely to, possess the proteases *splC splA*, *splB*, and *splD* than infectious isolates from Australia. Serine proteases are encoded on the νSaβ pathogenicity island [42,43]. The *spl* operon is present in most of *S. aureus* strains but some strains may not have the full operon [44]. Previous studies suggest that serine proteases are expressed during human infections and modulate *S. aureus* physiology and virulence [45], but their role in ocular infections is unknown. The current study suggests that some of these serine proteases may have a role in pathogenesis of conjunctivitis, and this should be studied in future experiments. Again, the trend of pathogenic isolates from USA infections to possess other serine proteases might be related to the different clonal types circulating in the USA. Studies reported ST5 (27%), ST8 (16%), ST30 (9%), and ST45 (6%) as prevalent clonal types in USA ocular isolates [46]. However, a study from tropical northern Australia reported CC75 as a prevalent clone in Australia [47]. In the current study 50% of infectious strains possessing serine proteases were CC5 (ST5, ST105, and ST840), 30% were CC8 (ST8), and 10% were CC15 (ST15) and CC30 (ST30). All strains from CC5 and CC30 possessed 4–6 proteases whereas strains from CC8 and CC15 possessed all six proteases. There was a greater spread of sequence types and clonal complexes in the Australian infectious isolates.

There were several differences in possession of toxins genes of the *set* family and others. The *set* genes are similar to staphylococcal superantigens but more likely to be involved in immune avoidance [48]. Several toxins in staphylococci are often carried on large mobile genetic elements (MGEs) known as pathogenicity islands that can be horizontally transferred [49] and can be located on the pathogenicity island SaPIn2, SaPII, and SaPIboy [50]. An increasing number of enterotoxins and enterotoxin-like genes in *S. aureus* have been identified and it is a global trend that around 80% of *S. aureus* both pathogenic and non-pathogenic isolates carry an average of 5–6 enterotoxin genes [51–53]. Whether the *set* genes in different strains were present on pathogenicity islands will be examined in future studies.

The current study's finding that enterotoxin E (*sea*) was more commonly found in ocular strains, infectious strains from USA (70%), AUS (33%), niCIE strains (50%) than non-ocular strains (20%), however, the current study findings are not consistent with earlier studies [54,55]. *S. aureus* strains isolated from atopic patients experiencing keratoconjunctivitis with corneal ulceration, possessed enterotoxins more frequently compared to patients with no ulceration [56]; the role of enterotoxins in ocular infections remains to be fully defined. Another study found enterotoxin and enterotoxin-like genes were found to be highly correlated with MRSA and predictive for MDR status in ocular isolates [57]. The antibiotic susceptibility data of the isolates used in the current study shows that most of the strains were multi-drug resistant (MDR) [12], so the distribution of enterotoxin-like genes in 28% of MRSA USA infectious strains (MK + conjunctivitis) may indicate their MDR status, but the presence of enterotoxin-like genes in MSSA (methicillin sensitive *S. aureus*) MDR strains from other conditions indicates enterotoxin genes are probably associated with the source of isolation. The genes *lukF-PV* and *lukS-PV* encode for the Panton–Valentine leukocidin which is linked to community acquired MRSA infections [58]. Their presence in conjunctivitis strains from USA (60%) in the current study supports previous studies which reported that Panton–Valentine leukocidin (*pvl*) is found in the majority of (67%) ocular strains [59].

The finding that there were no differences in the possession of genes related to immune evasion or type VII secretion systems (*adsA, chp, cpsA, scn, sbi, esaA, esaB, esaD, esaE, esaG, essA, essB, essC, esxA, esxB, esxC, esxD*) between different isolate types, countries, or ocular and non-ocular strains, and the finding that most isolates possessed the adhesin genes *atl, ebp, eap, efb, fnbA, fnbB, icaA, icaB, icaC, icaR, sdrC*, and *spa*, the enzyme genes (proteases, thermonuclease, lipase, staphylokinase, and hyaluronate lyase) *sspB, scpA, hysA, geh, lip, v8, sspA, sak*, and *nuc*, and the hemolysin genes *hla, hlgA, hlgB, hlgC* might indicate that possession of these genes is important either for survival in either the eye or in the environment prior to gaining access to the ocular surface to cause infection or inflammation.

The core and pan genome phylogenies included strains from all ocular conditions. Acquired genes are part of the pan rather than core genome [60] and the presence of larger pan genomes points towards the acquisition of new genes [61]. The core genome (which is almost 90% of pan or total genome) refers to the conserved genes present in a species, which might differ in each individual strain within that species [62]. With respect to multi-locus sequence typing, in the current study sequence types (STs), ST5 (20%), ST8 (16%), clonal complex (CC), CC30 (16%), and CC1 (12%) were the most prevalent (predominant) types respectively. Previous studies identified that specific lineages including ST5 and ST8 are common among *S. aureus* ocular strains [54,59,63–65]. ST5 MRSA isolates from USA were the frequent cause of hospital acquired infections [66] and ST8 MRSA isolates from USA most commonly the causes of community acquired skin and soft tissue infections [67,68]. This suggests that *S. aureus* isolates from ocular infections align with major circulating pathogenic *S. aureus* strains capable of causing systemic infections.

4. Material and Methods

4.1. Bacterial Isolates

Twenty-five *S. aureus* ocular isolates, 9 isolated from infections (MK + conjunctivitis) in Australia, 10 isolated from infections (MK + conjunctivitis) in USA and 6 isolated from niCIEs in Australia were used (Table 4). The isolates were selected from a larger collection of strains based on their published susceptibility to various antibiotics [12] and possession of virulence genes [13]. Most strains were multi-drug resistant (MDR; Table 4).

4.2. Whole Genome Sequencing

Genomic DNA from each *S. aureus* strain was extracted using QIAGEN DNeasy blood and tissue extraction kit (Hilden, North Rhine-Westphalia, Germany) as per the manufacturer's instructions. The Nextera XT DNA library preparation kit (Illumina, San Diego, CA, USA) was used to prepare paired-end libraries. All the libraries were multiplexed on one MiSeq run. FastQC version 0.117 (https://www.bioinformatics.babraham.ac.uk/projects/fastqc, accessed on 9 July 2021) was used to assess the quality of sequenced genomes using raw reads. Trimmomatic v0.38 (http://www.usadellab.org/cms/?page=trimmomatic, accessed on 9 July 2021) was used for trimming the adapters from the reads with the setting of minimum read length of 36 and minimum coverage of 15 [69]. De novo assembly using Spades v3.15.0 was performed using the default setting [70]. Assembled genomes were annotated with Prokka v1.12 using GeneBank® compliance flag [71]. The genome of *S. aureus* NCTC 8325 (reference strain in this study) was re-annotated with Prokka to avoid annotation bias.

Table 4. Susceptibility and virulence profiles of *S. aureus* strains [12,13].

Ocular Condition	Stain Number	Phenotypic Resistance (R) and Susceptibility (S) Profile	Profile of Virulence Genes Known to Be Possessed by the Isolates
Microbial keratitis USA	SA107	CIP, CEFT, OXA, AZI, POLYB (R) GN, VAN, CHL (S)	fnbpA, eap, sspB, sspA, coa, hla, hlg, hld.
	SA111	CIP, CEFT, OXA, GN, AZI, POLYB (R) VAN, CHL (S)	clfA, fnbpA, eap, sspB, sspA, coa, hla, hlg, hld
	SA112	CIP, CEFT, OXA, AZI, POLYB (R) GN, VAN, CHL (S)	clfA, fnbpA, eap, sspB, sspA, coa, hla, hlg, hld
	SA113	CIP, CEFT, OXA, AZI, POLYB (R) GN, VAN, CHL (S)	clfA, fnbpA, eap, sspB, sspA, coa, hla, hlg, hld
	SA114	CIP, CEFT, AZI, POLYB (R) GN, VAN, OXA, CHL (S)	clfA, fnbpA, eap, sspB, sspA, coa, hla, hlg, hld
Microbial keratitis Australia	SA34	CEFT, AZI, POLYB (R) CIP, GN, VAN, OXA, CHL (S)	fnbpA, eap, scpAsspB, sspA, coa, seb, hla, hlg, hld
	SA129	CEFT, CHL, AZI, POLYB (R) CIP, GN, VAN, OXA (S)	clfA, fnbpA, eap, sspA, coa, hla, hlg, hld
	M5-01	CIP, CEFT, CHL, AZI (R) GN, VAN, OXA, POLYB (S)	clfA, fnbpA, eap, scpA, sspB, sspA, coa, hla, hlg, hld
	M19-01	CEFT, AZI, POLYB (R) CIP, GN, VAN, OXA, CHL (S)	fnbpA, eap, scpA, sspB, sspA, coa, hla, hlg, hld
	M28-01	CEFT, CHL, AZI, POLYB (R) CIP, GN, VAN, OXA (S)	clfA, fnbpA, eap, scpA, sspB, sspA, coa, hla, hlg, hld
	M43-01	CIP, CEFT, OXA, CHL, AZI, POLYB (R) GN, VAN (S)	clfA, eap, scpA, sspB, sspA, coa, hla, hlg, hld
	M71-01	CIP, CEFT, CHL, AZI, POLYB (R) GN, VAN, OXA (S)	clfA, fnbpA, eap, scpA, sspB, sspA, coa, hla, hlg, hld
Conjunctivitis USA	SA86	CEFT, CHL, AZI, POLYB (R) CIP, GN, VAN, OXA (S)	clfA, fnbpA, eap, scpA, sspB, coa, seb, hla, hlg, hld, pvl.
	SA90	CIP, CEFT, AZI, POLYB (R) GN, VAN, OXA, CHL (S)	clfA, fnbpA, eap, scpA, sspB, coa, hla, hlg, hld.
	SA101	CIP, CEFT, OXA, AZI, POLYB (R) GN, VAN, CHL (S)	clfA, fnbpA, eap, sspB, sspA, coa, hla, hlg, hld, pvl
	SA102	CIP, CEFT, OXA, AZI, POLYB (R) GN, VAN, CHL (S)	clfA, fnbpA, eap, sspB, sspA, coa, seb, hla, hlg, hld.
	SA103	CIP, CEFT, OXA, AZI, POLYB (R) GN, VAN, CHL (S)	clfA, fnbpA, eap, sspB, sspA, coa, hla, hlg, hld, pvl
Conjunctivitis Australia	SA46	AZI, POLYB (R) CIP, CEFT, OXA, GN, VAN, CHL (S)	clfA, fnbpA, eap, scpA, sspB, sspA, coa, hla, hlg, hld.
	SA136	CIP, CEFT, AZI, POLYB (R) GN, VAN, OXA, CHL (S)	fnbpA, eap, scpA, sspB, sspA, coa, hla, hlg, hld.
niCIE Australia	SA20	CEFT, CHL, AZI, POLYB (R) CIP, GN, VAN, OXA (S)	fnbpA, eap, scpA, sspB, sspA, coa, seb, hla, hlg, pvl.
	SA25	AZI, POLYB (R) CIP, CEFT, GN, VAN, OXA, CHL (S)	clfA, fnbpA, eap, scpA, sspB, sspA, coa, seb, hla, hlg, hld.
	SA27	CEFT, OXA, AZI, POLYB (R) CIP, GN, VAN, CHL (S)	clfA, fnbpA, eap, scpA, sspB, sspA, coa, hla, hlg, hld.
	SA31	CIP, CEFT, AZI, POLYB (R) GN, VAN, OXA, CHL (S)	clfA, fnbpA, eap, scpA, sspB, sspA, coa, hla, hlg, hld.
	SA32	POLYB (R) CIP, CEFT, AZI, GN, VAN, OXA, CHL (S)	clfA, fnbpA, eap, scpA, sspB, sspA, coa, hla, hlg, hld.
	SA48	CEFT, CHL, AZI, POLYB (R) CIP, GN, VAN, OXA (S).	clfA, fnbpA, eap, sspB, sspA, coa, hla, hlg.

R = resistant, S = sensitive; CIP = ciprofloxacin, CEFT = ceftazidime, OXA = oxacillin, GN = gentamicin, VAN = vancomycin, CHL = chloramphenicol, AZI = azithromycin, POLYB = polymyxin B. *clfA* = clumping factor, *fnbpA* = fibronectin binding protein, *eap* = extracellular adhesion protein, *scpA* = cysteine protease staphopain A, *sspB* = cysteine protease staphopain B, *sspA* = serine protease v8, *coa* = collagen binding adhesion, *seb* = enterotoxin, *hla* = alpha-toxin, *hlg* = gamma-toxin, *hld* = delta-toxin, *pvl* = Panton–Valentine leukocidin.

Multi-locus sequence type (MLST) was determined using PubMLST (https://pubmlst.org/, accessed on 29 September 2021) [72] to find the sequence of each strain. Pan genomes of the *S. aureus* isolates were analyzed using Roary v3.11.2 [73] which uses the GFF3 files produced by Prokka. The program was run using the default settings, which uses BLASTp for all-against-all comparison with 95% of percentage sequence identity. Core genes were taken as the genes which were common in at least 99% of strains. Core genome phylogeny was constructed using Harvest Suite Parsnp v1.2 [74] with *S. aureus* NCTC

8325 (NC_007795.1) as a reference strain. The output file 'genes_ presence_absence.csv' generated by Roary was used to compare the *S. aureus* isolates. Phylogenetic tree was constructed using online webtool itol (https://itol.embl.de/, accessed on 4 April 2022). Acquired antibiotic resistance genes of *S. aureus* isolates were examined by using the online database Resfinder v3.1 (https://cge.cbs.dtu.dk/services/ResFinder/, accessed on 30 October 2021) [75]. To determine the association of specific virulence determinants with specific ocular conditions, 128 virulence factors previously described to be associated with many *S. aureus* infections in the virulence factors database (VFDB) were examined (VFDB; Centre for Genomic Epidemiology, DTU, Denmark, http://www.mgc.ac.cn/VFs/main.htm, (accessed on 9 May 2022) respectively [76]. The assembled *S. aureus* isolates were compared to a custom VFDB consisting of 128 virulence genes associated with adhesion, enzymes, immune evasion, type VII secretion systems, and toxins (enterotoxins, enterotoxin-like genes, exfoliative toxins, and exotoxins). A gene sequence had to cover at least 60% of the length of the gene sequence in the database with a sequence identity of 90% to be considered as being present in the strain. As acquired antimicrobial resistance genes may be carried on integrons, *S. aureus* genomes were analyzed for integrons using integron Finder version 1.5.1 (https://bioweb.pasteur.fr/packages/pack@Integron_Finder@1.5.1, accessed on 5 February 2022). There was no evidence for integrons in these 25 isolates. Isolates with the same sequence types were compared for nucleotide similarities using the MUMmer online web tool (http://jspecies.ribohost.com/jspeciesws/#analyse, accessed on 18 February 2022).

4.3. Statistical Analysis

Differences in virulence factors database (VFDB) results for the presence or absence of virulence genes between the disease groups and differences in ocular and non-ocular isolates were analyzed using a chi-square test in GraphPad prism v8.0.2.263 for windows (San Diego, CA, USA, (www.graphpad.com, accessed on 1 June 2022). For all analyses a p-value < 0.05 was considered statistically significant and p-value < 0.1 was considered as trending towards significance.

Nucleotide accession: The nucleotide sequences are available in the Genebank under the Bio project accession number PRJNA859391 (https://www.ncbi.nlm.nih.gov/bioproject/PRJNA859391, accessed on 24 July 2022), (genomes accession number (JANHMY000000000, JANHMZ000000000, JANHNA000000000, JANHNB000000000, JANHNC000000000, JANHND000000000, JANHNE000000000, JANHNF000000000, JANHNG000000000, JANHNH000000000, JANHNI000000000, JANHNJ000000000, JANHNK000000000, JANHNL000000000, JANHNM000000000, JANHNN000000000, JANHNO000000000, JANHNP000000000, JANHNQ000000000, JANHNR000000000, JANHNS000000000, JANHNT000000000, JANHNU000000000, JANHNV000000000, JANHNW000000000).

5. Conclusions

With respect to virulence determinants distribution, there were some differences between ocular and non-ocular isolates and ocular infectious and niCIE isolates. The current study could not detect plasmids in any of the isolates, as it relied on draft genomes. Further studies including more strains will focus on improvement of the assembly and probe the WGS for possession of pathogenicity islands such as vSaβ, SaPIn2, SaPII, and SaPIboy. Overall, these findings have extended our understanding of the genomic diversity of *S. aureus* in infectious and non-infectious ocular conditions. The information can be used to elucidate various mechanisms that would help combat virulent and drug resistant strains.

Supplementary Materials: The following supporting information can be downloaded at: https://www.mdpi.com/article/10.3390/antibiotics11081011/s1, Table S1: Genomic features of *S. aureus* non-ocular isolates.

Author Contributions: Conceptualization, M.A., M.D.P.W., F.S. and A.K.V.; methodology, M.A., M.D.P.W., F.S. and A.K.V.; writing—original draft preparation, M.A.; writing—review and editing, M.D.P.W., F.S. and A.K.V.; supervision, M.D.P.W., F.S. and A.K.V.; funding acquisition, M.D.P.W. All authors have read and agreed to the published version of the manuscript.

Funding: This research received no external funding.

Institutional Review Board Statement: *S. aureus* strains from the Bascom Palmer Institute, Miami (USA), were kindly provided by Dr. Darlene Miller. All strains were donated without identifiable patient data and ethics was not required to send or receive these strains. The more recent strains from the Prince of Wales Hospital (Australia) were kindly provided by Dr. Monica Lahra. Ethics was obtained (HREA Application ID: 2020/ETH02783), approval was required and obtained for these strains to be transferred to the School of Optometry and Vision Science, UNSW. No clinical data was provided that could be identified back to the patients from which the bacterial strains had been collected.

Informed Consent Statement: Not Applicable.

Data Availability Statement: Data is contained within the article and available upon request.

Acknowledgments: The authors would like to acknowledge Darlene Miller, Bascom Palmer Institute, Miami (USA) and Monica Lahra, Prince of Wales Hospital Sydney, for providing *S. aureus* MK strains. The authors would also like to acknowledge Associate Scott Rice and Stephen Summers and genome facility of the Singapore Centre of Life Science Engineering, Nanyang Technological University, Singapore for providing the sequencing. We are also thankful to UNSW high performance computing facility KATANA for providing us with the cluster time for the data analysis.

Conflicts of Interest: The authors declare no conflict of interest.

References

1. Mainous, A.G., III; Hueston, W.J.; Everett, C.J.; Diaz, V.A. Nasal carriage of *Staphylococcus aureus* and methicillin-resistant *S. aureus* in the United States, 2001–2002. *Ann. Fam. Med.* **2006**, *4*, 132–137. [CrossRef] [PubMed]
2. Collier, S.A.; Gronostaj, M.P.; MacGurn, A.K.; Cope, J.R.; Awsumb, K.L.; Yoder, J.S.; Beach, M.J. Estimated burden of keratitis-United States, 2010. *Morb. Mortal. Wkly. Rep.* **2014**, *63*, 1027–1030.
3. Shields, T.; Sloane, P.D. A comparison of eye problems in primary care and ophthalmology practices. *Fam. Med.* **1991**, *23*, 544–546. [PubMed]
4. Green, M.; Carnt, N.; Apel, A.; Stapleton, F. Queensland Microbial Keratitis Database: 2005-2015. *Br. J. Ophthalmol.* **2019**, *103*, 1481–1486. [CrossRef]
5. Mah, F.S.; Davidson, R.; Holland, E.J.; Hovanesian, J.; John, T.; Kanellopoulos, J.; Kim, T. Current knowledge about and recommendations for ocular methicillin-resistant *Staphylococcus aureus*. *J. Cataract. Refract. Surg.* **2014**, *40*, 1894–1908. [CrossRef] [PubMed]
6. Jin, H.; Parker, W.T.; Law, N.W.; Clarke, C.L.; Gisseman, J.D.; Pflugfelder, S.C.; Al-Mohtaseb, Z.N. Evolving risk factors and antibiotic sensitivity patterns for microbial keratitis at a large county hospital. *Br. J. Ophthalmol.* **2017**, *101*, 1483–1487. [CrossRef]
7. Sand, D.; She, R.; Shulman, I.A.; Chen, D.S.; Schur, M.; Hsu, H.Y. Microbial keratitis in los angeles: The doheny eye institute and the los angeles county hospital experience. *Ophthalmology* **2015**, *122*, 918–924. [CrossRef]
8. Wong, V.W.; Lai, T.Y.; Chi, S.C.; Lam, D.S. Pediatric ocular surface infections: A 5-year review of demographics, clinical features, risk factors, microbiological results, and treatment. *Cornea* **2011**, *30*, 995–1002. [CrossRef] [PubMed]
9. Sweeney, D.F.; Jalbert, I.; Covey, M.; Sankaridurg, P.R.; Vajdic, C.; Holden, B.A.; Rao, G.N. Clinical characterization of corneal infiltrative events observed with soft contact lens wear. *Cornea* **2003**, *22*, 435–442. [CrossRef] [PubMed]
10. Otto, M. Basis of virulence in community-associated methicillin-resistant *Staphylococcus aureus*. *Ann. Rev. Microbiol.* **2010**, *64*, 143–162. [CrossRef] [PubMed]
11. Cheung, G.Y.C.; Bae, J.S.; Otto, M. Pathogenicity and virulence of *Staphylococcus aureus*. *Virulence* **2021**, *12*, 547–569. [CrossRef] [PubMed]
12. Afzal, M.; Vijay, A.K.; Stapleton, F.; Willcox, M. Susceptibilty of ocular *Staphylococcus aureus* to antibioticsand multipurpose disinfecting soutions. *Antibiotics* **2021**, *10*, 1203. [CrossRef] [PubMed]
13. Afzal, M.; Vijay, A.K.; Stapleton, F.; Willcox, M. Virulence genes of *Staphylococcus aureus* associated with keratitis, conjunctivitis and contact lens-assocaied inflammation. *Transl. Vis. Sci. Technol.* **2022**, *11*, 5. [CrossRef] [PubMed]
14. Humphreys, H.; Coleman, D.C. Contribution of whole-genome sequencing to understanding of the epidemiology and control of meticillin-resistant *Staphylococcus aureus*. *J. Hosp. Infect.* **2019**, *102*, 189–199. [CrossRef]
15. Leopold, S.R.; Goering, R.V.; Witten, A.; Harmsen, D.; Mellmann, A. Bacterial whole-genome sequencing revisited: Portable, scalable, and standardized analysis for typing and detection of virulence and antibiotic resistance genes. *J. Clin. Microbiol.* **2014**, *52*, 2365–2370. [CrossRef] [PubMed]

16. Recker, M.; Laabei, M.; Toleman, M.S.; Reuter, S.; Saunderson, R.B.; Blane, B.; Massey, R.C. Clonal differences in *Staphylococcus aureus* bacteraemia-associated mortality. *Nat. Microbiol.* **2017**, *2*, 1381–1388. [CrossRef]
17. Lilje, B.; Rasmussen, R.V.; Dahl, A.; Stegger, M.; Skov, R.L.; Fowler, V.G.; Ng, K.L.; Kiil, K.; Larsen, A.R.; Petersen, A.; et al. Whole-genome sequencing of bloodstream Staphylococcus aureus isolates does not distinguish bacteraemia from endocarditis. *Microb. Genom.* **2017**, *3*, e000138. [CrossRef]
18. Manara, S.; Pasolli, E.; Dolce, D.; Ravenni, N.; Campana, S.; Armanini, F.; Asnicar, F.; Mengoni, A.; Galli, L.; Montagnani, C.; et al. Whole-genome epidemiology, characterisation, and phylogenetic reconstruction of Staphylococcus aureus strains in a paediatric hospital. *Genome Med.* **2018**, *10*, 82. [CrossRef]
19. Wildeman, P.; Tevell, S.; Eriksson, C.; Lagos, A.C.; Söderquist, B.; Stenmark, B. Genomic characterization and outcome of prosthetic joint infections caused by *Staphylococcus aureus*. *Sci. Rep.* **2020**, *10*, 5938. [CrossRef] [PubMed]
20. Asbell, P.A.; DeCory, H.H. Antibiotic resistance among bacterial conjunctival pathogens collected in the antibiotic resistance monitoring in ocular microorganisms (ARMOR) surveillance study. *PLoS ONE* **2018**, *13*, e0205814. [CrossRef] [PubMed]
21. Diekema, D.J.; Pfaller, M.A.; Schmitz, F.J.; Smayevsky, J.; Bell, J.; Jones, R.N.; Beach, M.; SENTRY Partcipants Group. Survey of infections due to *Staphylococcus species*: Frequency of occurrence and antimicrobial susceptibility of isolates collected in the United States, Canada, Latin America, Europe, and the Western Pacific region for the SENTRY antimicrobial surveillance program (SARP), 1997–1999. *Clin. Infect. Dis.* **2001**, *32*, 114–132.
22. Cabrera-Aguas, M.; Khoo, P.; George, C.R.R.; Lahra, M.M.; Watson, S.L. Antimicrobial resistance trends in bacterial keratitis over 5 years in Sydney, Australia. *Clin. Exp. Ophthalmol.* **2020**, *48*, 183–191. [CrossRef] [PubMed]
23. Coombs, G.W.; Bell, J.M.; Collignon, P.J.; Nimmo, G.R.; Christiansen, K.J. Prevalence of MRSA strains among *Staphylococcus aureus* isolated from outpatients, 2006. Report from the Australian Group for Antimicrobial Resistance. *Commun. Dis. Intell. Q. Rep.* **2009**, *33*, 10–20.
24. Kwiecinski, J.; Jin, T.; Josefsson, E. Surface proteins of *Staphylococcus aureus* play an important role in experimental skin infection. *Apmis* **2014**, *122*, 1240–1250. [CrossRef] [PubMed]
25. Sundin, G.W.; Bender, C.L. Dissemination of the strA-strB streptomycin-resistance genes among commensal and pathogenic bacteria from humans, animals, and plants. *Mol. Ecol.* **1996**, *5*, 133–143. [CrossRef]
26. Khosravi, A.D.; Jenabi, A.; Montazeri, E.A. Distribution of genes encoding resistance to aminoglycoside modifying enzymes in methicillin-resistant *Staphylococcus aureus* (MRSA) strains. *Kaohsiung J. Med. Sci.* **2017**, *33*, 587–593. [CrossRef]
27. Yadegar, A.; Sattari, M.; Mozafari, N.A.; Goudarzi, G.R. Prevalence of the genes encoding aminoglycoside-modifying enzymes and methicillin resistance among clinical isolates of *Staphylococcus aureus* in Tehran, Iran. *Microb. Drug Resist.* **2009**, *15*, 109–113. [CrossRef]
28. Ramirez, M.S.; Tolmasky, M.E. Aminoglycoside modifying enzymes. *Drug Resist. Updates* **2010**, *13*, 151–171. [CrossRef]
29. Gad, G.F.; El-Feky, M.A.; El-Rehewy, M.S.; Hassan, M.A.; Abolella, H.; El-Baky, R.M. Detection of icaA, icaD genes and biofilm production by *Staphylococcus aureus* and *Staphylococcus epidermidis* isolated from urinary tract catheterized patients. *J. Infect. Dev. Ctries.* **2009**, *3*, 342–351.
30. Willcox, M.D.P.; Carnt, N.; Diec, J.; Naduvilath, T.; Evans, V.; Stapleton, F.; Holden, B.A. Contact lens case contamination during daily wear of silicone hydrogels. *Optom. Vis. Sci.* **2010**, *87*, 456–464. [CrossRef]
31. Rhem, M.N.; Lech, E.M.; Patti, J.M.; McDevitt, D.; Hook, M.; Jones, D.B.; Wilhelmus, K.R. The collagen-binding adhesin is a virulence factor in *Staphylococcus aureus* keratitis. *Infec. Immun.* **2000**, *68*, 3776–3779. [CrossRef]
32. Cerca, N.; Brooks, J.L.; Jefferson, K.K. Regulation of the intercellular adhesin locus regulator (icaR) by SarA, sigmaB, and IcaR in *Staphylococcus aureus*. *J. Bacteriol.* **2008**, *190*, 6530–6533. [CrossRef] [PubMed]
33. Jefferson, K.K.; Pier, D.B.; Goldmann, D.A.; Pier, G.B. The teicoplanin-associated locus regulator (TcaR) and the intercellular adhesin locus regulator (IcaR) are transcriptional inhibitors of the locus in *Staphylococcus aureus*. *J. Bacteriol.* **2004**, *186*, 2449–2456. [CrossRef] [PubMed]
34. Clarke, S.R.; Harris, L.G.; Richards, R.G.; Foster, S.J. Analysis of *Ebh*, a 1.1-megadalton cell wall-associated fibronectin-binding protein of *Staphylococcus aureus*. *Infect. Immu.* **2002**, *70*, 6680–6687. [CrossRef]
35. Cheng, A.G.; Missiakas, D.; Schneewind, O. The giant protein *Ebh* is a determinant of *Staphylococcus aureus* cell size and complement resistance. *J. Bacteriol.* **2014**, *196*, 971–981. [CrossRef] [PubMed]
36. Kuroda, M.; Tanaka, Y.; Aoki, R.; Shu, D.; Tsumoto, K.; Ohta, T. *Staphylococcus aureus* giant protein Ebh is involved in tolerance to transient hyperosmotic pressure. *Biochem. Biophys. Res. Commun.* **2008**, *374*, 237–241. [CrossRef]
37. Foster, T.J.; Geoghegan, J.A.; Ganesh, V.K.; Höök, M. Adhesion, invasion and evasion: The many functions of the surface proteins of *Staphylococcus aureus*. *Nat. Rev. Microb.* **2014**, *12*, 49–62. [CrossRef]
38. Corrigan, R.M.; Miajlovic, H.; Foster, T.J. Surface proteins that promote adherence of *Staphylococcus aureus* to human desquamated nasal epithelial cells. *BMC Microbiol.* **2009**, *9*, 22. [CrossRef]
39. Askarian, F.; Ajayi, C.; Hanssen, A.M.; Van Sorge, N.M.; Pettersen, I.; Diep, D.B.; Johannessen, M. The interaction between *Staphylococcus aureus SdrD* and desmoglein 1 is important for adhesion to host cells. *Sci. Rep.* **2016**, *6*, 22134. [CrossRef]
40. Cheng, A.G.; Kim, H.K.; Burts, M.L.; Krausz, T.; Schneewind, O.; Missiakas, D.M. Genetic requirements for *Staphylococcus aureus* abscess formation and persistence in host tissues. *FASEB J.* **2009**, *23*, 3393–3404. [CrossRef]

41. Sabat, A.; Melles, D.C.; Martirosian, G.; Grundmann, H.; van Belkum, A.; Hryniewicz, W. Distribution of the serine-aspartate repeat protein-encoding *sdr* genes among nasal-carriage and invasive *Staphylococcus aureus* strains. *J. Clin. Microbio.* 2006, *44*, 1135–1138. [CrossRef]
42. Baba, T.; Bae, T.; Schneewind, O.; Takeuchi, F.; Hiramatsu, K. Genome sequence of *Staphylococcus aureus* strain newman and comparative analysis of *Staphylococcal* genomes: Polymorphism and evolution of two major pathogenicity islands. *J. Bacteriol.* 2008, *190*, 300–310. [CrossRef]
43. Zdzalik, M.; Karim, A.Y.; Wolski, K.; Buda, P.; Wojcik, K.; Brueggemann, S.; Jonsson, I.M. Prevalence of genes encoding extracellular proteases in *Staphylococcus aureus*—Important targets triggering immune response in vivo. *FEMS Immunol. Med. Microbiol.* 2012, *66*, 220–229. [CrossRef] [PubMed]
44. Reed, S.B.; Wesson, C.A.; Liou, L.E.; Trumble, W.R.; Schlievert, P.M.; Bohach, G.A.; Bayles, K.W. Molecular characterization of a novel *Staphylococcus aureus* serine protease operon. *Infect. Immun.* 2001, *69*, 1521–1527. [CrossRef] [PubMed]
45. Paharik, A.E.; Salgado-Pabon, W.; Meyerholz, D.K.; White, M.J.; Schlievert, P.M.; Horswill, A.R. The *Spl* Serine proteases modulate *Staphylococcus aureus* protein production and virulence in a rabbit model of pneumonia. *mSphere* 2016, *1*, e00208–e00216. [CrossRef]
46. Johnson, W.L.; Sohn, M.B.; Taffner, S.; Chatterjee, P.; Dunman, P.M.; Pecora, N.; Wozniak, R.A.F. Genomics of *Staphylococcus aureus* ocular isolates. *PLoS ONE* 2021, *16*, e0250975. [CrossRef]
47. Ng, J.W.S.; Holt, D.C.; Lilliebridge, R.A.; Stephens, A.J.; Huygens, F.; Tong, S.Y.C.; Currie, B.J.; Giffard, P.M. Phylogenetically Distinct *Staphylococcus aureus* Lineage Prevalent among Indigenous Communities in Northern Australia. *J. Clin. Microbiol.* 2009, *47*, 2295–2300. [CrossRef] [PubMed]
48. Bretl, D.J.; Elfessi, A.; Watkins, H.; Schwan, W.R. Regulation of the staphylococcal superantigen-like protein 1 gene of community-associated methicillin-resistant *Staphylococcus aureus* in murine abscesses. *Toxins* 2019, *11*, 391. [CrossRef] [PubMed]
49. Grumann, D.; Nübel, U.; Bröker, B.M. *Staphylococcus aureus* toxins–Their functions and genetics. *Infec. Gen. Evol.* 2014, *21*, 583–592. [CrossRef]
50. Kuroda, M.; Ohta, T.; Uchiyama, I.; Baba, T.; Yuzawa, H.; Kobayashi, I.; Hiramatsu, K. Whole genome sequencing of meticillin-resistant *Staphylococcus aureus*. *Lancet* 2001, *357*, 1225–1240. [CrossRef]
51. Becker, K.; Friedrich, A.W.; Peters, G.; von Eiff, C. Systematic survey on the prevalence of genes coding for staphylococcal enterotoxins SElM, SElO, and SElN. *Mol. Nut. Food Res.* 2004, *48*, 488–495. [CrossRef] [PubMed]
52. Fisher, E.L.; Otto, M.; Cheung, G.Y.C. Basis of virulence in enterotoxin-mediated *staphylococcal* food poisoning. *Front. Microbiol.* 2018, *9*, 436. [CrossRef] [PubMed]
53. Jarraud, S.; Peyrat, M.A.; Lim, A.; Tristan, A.; Bes, M.; Mougel, C.; Lina, G. A highly prevalent operon of enterotoxin gene forms a putative nursery of superantigens in *Staphylococcus aureus*. *J. Immunol.* 2001, *166*, 669–677. [CrossRef]
54. Peterson, J.C.; Durkee, H.; Miller, D.; Maestre-Mesa, J.; Arboleda, A.; Aguilar, M.C.; Alfonso, E. Molecular epidemiology and resistance profiles among healthcare and community-associated *Staphylococcus aureus* keratitis isolates. *Infect. Drug Resist.* 2019, *12*, 831–843. [CrossRef] [PubMed]
55. Kłos, M.; Pomorska-Wesołowska, M.; Romaniszyn, D.; Chmielarczyk, A.; Wójkowska-Mach, J. Epidemiology, drug resistance, and virulence of *Staphylococcus aureus* isolated from ocular infections in polish patients. *Pol. J. Microbiol.* 2019, *68*, 541–548. [CrossRef]
56. Fujishima, H.; Okada, N.; Dogru, M.; Baba, F.; Tomita, M.; Abe, J.; Saito, H. The role of *Staphylococcal* enterotoxin in atopic keratoconjunctivitis and corneal ulceration. *Allergy* 2012, *67*, 799–803. [CrossRef]
57. Lu, M.; Parel, J.M.; Miller, D. Interactions between *staphylococcal* enterotoxins A and D and superantigen-like proteins 1 and 5 for predicting methicillin and multidrug resistance profiles among *Staphylococcus aureus* ocular isolates. *PLoS ONE* 2021, *16*, e0254519. [CrossRef]
58. Vandenesch, F.; Naimi, T.; Enright, M.C.; Lina, G.; Nimmo, G.R.; Heffernan, H.; Etienne, J. Community-acquired methicillin-resistant *Staphylococcus aureus* carrying Panton-Valentine leukocidin genes: Worldwide emergence. *Emerg. Infect. Dis.* 2003, *9*, 978–984. [CrossRef]
59. Kang, Y.C.; Hsiao, C.H.; Yeh, L.K.; Ma, D.H.K.; Chen, P.Y.F.; Lin, H.C.; Huang, Y.C. Methicillin-resistant *Staphylococcus aureus* ocular infection in Taiwan: Clinical features, genotying, and antibiotic susceptibility. *Medicine* 2015, *94*, e1620. [CrossRef]
60. Roy, P.H.; Tetu, S.; Larouche, A.; Elbourne, L.; Tremblay, S.; Ren, Q.; Dodson, R.; Harkins, D.; Shay, R.; Watkins, K.; et al. Complete genome sequence of the multiresistant taxonomic outlier *Pseudomonas aeruginosa* PA7. *PLoS ONE* 2010, *5*, e8842. [CrossRef]
61. Subedi, D.; Vijay, A.K.; Kohli, G.S.; Rice, S.A.; Willcox, M. Comparative genomics of clinical strains of *Pseudomonas aeruginosa* strains isolated from different geographic sites. *Sci. Rep.* 2018, *8*, 15668. [CrossRef] [PubMed]
62. Wolfgang, M.C.; Kulasekara, B.R.; Liang, X.; Boyd, D.; Wu, K.; Yang, Q.; Lory, S. Conservation of genome content and virulence determinants among clinical and environmental isolates of *Pseudomonas aeruginosa*. *Proc. Natl. Acad. Sci. USA* 2003, *100*, 8484–8489. [CrossRef]
63. Nithya, V.; Rathinam, S.; Siva Ganesa Karthikeyan, R.; Lalitha, P. A ten-year study of prevalence, antimicrobial susceptibility pattern, and genotypic characterization of Methicillin resistant *Staphylococcus aureus* causing ocular infections in a tertiary eye care hospital in South India. *Infect. Genet. Evol.* 2019, *69*, 203–210. [CrossRef]
64. Wurster, J.I.; Bispo, P.J.M.; Van Tyne, D.; Cadorette, J.J.; Boody, R.; Gilmore, M.S. *Staphylococcus aureus* from ocular and otolaryngology infections are frequently resistant to clinically important antibiotics and are associated with lineages of community and hospital origins. *PLoS ONE* 2018, *13*, e0208518. [CrossRef] [PubMed]

65. Carrel, M.; Perencevich, E.N.; David, M.Z. USA300 methicillin-resistant *Staphylococcus aureus*, United States, 2000-2013. *Emerg. Infect. Dis.* **2015**, *21*, 1973–1980. [CrossRef]
66. Challagundla, L.; Reyes, J.; Rafiqullah, I.; Sordelli, D.O.; Echaniz-Aviles, G.; Velazquez-Meza, M.E.; Robinson, D.A. Phylogenomic classification and the evolution of clonal complex 5 methicillin-esistant *Staphylococcus aureus* in the Western Hemisphere. *Front. Microbiol.* **2018**, *9*, 10. [CrossRef] [PubMed]
67. Strauß, L.; Stegger, M.; Akpaka, P.E.; Alabi, A.; Breurec, S.; Coombs, G.; Mellmann, A. Origin, evolution, and global transmission of community-acquired *Staphylococcus aureus* ST8. *Proc. Natl. Acad. Sci. USA* **2017**, *114*, e10596–e10604. [CrossRef]
68. Tenover, F.C.; Goering, R.V. Methicillin-resistant *Staphylococcus aureus* strain USA300: Origin and epidemiology. *J. Antimicrob. Chemother.* **2009**, *64*, 441–446. [CrossRef]
69. Bolger, A.M.; Lohse, M.; Usadel, B. Trimmomatic: A flexible trimmer for Illumina sequence data. *Bioinformatics* **2014**, *30*, 2114–2120. [CrossRef]
70. Nurk, S.; Bankevich, A.; Antipov, D.; Gurevich, A.; Korobeynikov, A.; Lapidus, A.; Sirotkin, Y. Assembling genomes and mini-metagenomes from highly chimeric reads. In *Annual International Conference on Research in Computational Molecular Biology*; Springer: Berlin/Heidelberg, Germany, 2013; pp. 158–170.
71. Seemann, T. Prokka: Rapid prokaryotic genome annotation. *Bioinformatics* **2014**, *30*, 2068–2069. [CrossRef]
72. Jolley, K.A.; Maiden, M.C.J. BIGSdb: Scalable analysis of bacterial genome variation at the population level. *BMC Bioinform.* **2010**, *11*, 595. [CrossRef]
73. Page, A.J.; Cummins, C.A.; Hunt, M.; Wong, V.K.; Reuter, S.; Holden, M.T.; Parkhill, J. Roary: Rapid large-scale prokaryote pan genome analysis. *Bioinformatics* **2015**, *31*, 3691–3693. [CrossRef] [PubMed]
74. Treangen, T.J.; Ondov, B.D.; Koren, S.; Phillippy, A.M. The Harvest suite for rapid core-genome alignment and visualization of thousands of intraspecific microbial genomes. *Genome Biol.* **2014**, *15*, 524. [CrossRef] [PubMed]
75. Zankari, E.; Hasman, H.; Cosentino, S.; Vestergaard, M.; Rasmussen, S.; Lund, O.; Larsen, M.V. Identification of acquired antimicrobial resistance genes. *J. Antimicrob. Chemother.* **2012**, *67*, 2640–2644. [CrossRef] [PubMed]
76. Joensen, K.G.; Scheutz, F.; Lund, O.; Hasman, H.; Kaas, R.S.; Nielsen, E.M.; Aarestrup, F.M. Real-time whole-genome sequencing for routine typing, surveillance, and outbreak detection of verotoxigenic *Escherichia coli*. *J. Clinic. Microbiol.* **2014**, *52*, 1501–1510. [CrossRef] [PubMed]

Article

Clinical Features and Molecular Characteristics of Methicillin-Susceptible *Staphylococcus aureus* Ocular Infection in Taiwan

Yueh-Ling Chen [1], Eugene Yu-Chuan Kang [1,2], Lung-Kun Yeh [1,2], David H. K. Ma [1,2], Hsin-Yuan Tan [1,2], Hung-Chi Chen [1,2], Kuo-Hsuan Hung [1,2], Yhu-Chering Huang [2,3] and Ching-Hsi Hsiao [1,2,*]

1 Department of Ophthalmology, Chang Gung Memorial Hospital, Taoyuan 333, Taiwan; yuehling20@gmail.com (Y.-L.C.); yckang0321@gmail.com (E.Y.-C.K.); lkyeh@ms9.hinet.net (L.-K.Y.); davidhkma@yahoo.com (D.H.K.M.); tanhsin@gmail.com (H.-Y.T.); mr3756@cgmh.org.tw (H.-C.C.); agarlic2000@gmail.com (K.-H.H.)
2 College of Medicine, Chang Gung University, Taoyuan 333, Taiwan; ychuang@cgmh.org.tw
3 Division of Pediatric Infectious Diseases, Department of Pediatrics, Chang Gung Memorial Hospital, Taoyuan 333, Taiwan
* Correspondence: hsiao.chinghsi@gmail.com; Tel.: +886-3-3281200

Abstract: This study analyzed the clinical features and molecular characteristics of methicillin-susceptible *Staphylococcus aureus* (MSSA) ocular infections in Taiwan and compared them between community-associated (CA) and health-care-associated (HA) infections. We collected *S. aureus* ocular isolates from patients at Chang Gung Memorial Hospital between 2010 and 2017. The infections were classified as CA or HA using epidemiological criteria, and the isolates were molecularly characterized using pulsed-field gel electrophoresis, multilocus sequence typing, and Panton-Valentine leukocidin (PVL) gene detection. Antibiotic susceptibility was evaluated using disk diffusion and an E test. A total of 104 MSSA ocular isolates were identified; 46 (44.2%) were CA-MSSA and 58 (55.8%) were HA-MSSA. Compared with HA-MSSA strains, CA-MSSA strains caused a significantly higher rate of keratitis, but a lower rate of conjunctivitis. We identified 14 pulsotypes. ST 7/pulsotype BA was frequently identified in both CA-MSSA (28.3%) and HA-MSSA (37.9%) cases. PVL genes were identified in seven isolates (6.7%). Both CA-MSSA and HA-MSSA isolates were highly susceptible to vancomycin, teicoplanin, tigecycline, sulfamethoxazole–trimethoprim, and fluoroquinolones. The most common ocular manifestations were keratitis and conjunctivitis for CA-MSSA and HA-MSSA, respectively. The MSSA ocular isolates had diverse molecular characteristics; no specific genotype differentiated CA-MSSA from HA-MSSA. Both strains exhibited similar antibiotic susceptibility.

Keywords: *Staphyloccus aureus*; MSSA; ocular infection; pulsed-field gel electrophoresis; multilocus sequence typing; Panton-Valentine leukocidin; antibiotic resistance

1. Introduction

Staphylococcus aureus is a major isolated bacterial pathogen that causes various infections in humans [1]. *S. aureus* is typically categorized as methicillin resistant or methicillin susceptible based on its susceptibility to methicillin. Methicillin-resistant *S. aureus* (MRSA), a strain resistant to all β-lactam antibiotics, warrants particular attention because of its potentially limited treatment options and its increasing prevalence [2]. MRSA has conventionally been considered a health-care-associated (HA) pathogen but has been increasingly reported in community-associated (CA) infections, which is a global health concern [3]. HA- and CA-MRSA strains exhibit distinct clinical presentations, genotypes, and phenotypes [2,4].

Although MRSA has received considerable research attention, methicillin-susceptible *S. aureus* (MSSA) infections are more prevalent than MRSA infections [5]. According to surveillance in eight United States (US) counties in 2016, the incidence of invasive MSSA was 1.8 times higher than that of MRSA; MSSA accounted for 59.7% of HA cases and 60.1% of deaths [5]. In addition, a report published by the Centers for Disease Control and Prevention (CDC) revealed that the incidence of MSSA bacteremia increased by 3.9% annually from 2012 to 2017 in the community in the US [6]. However, few studies have explored the effect of healthcare exposure on the clinical features and molecular typing of MSSA infections [7–10], potentially due to the clonal diversity of MSSA infections. Most MRSA clones are considered to have evolved from epidemic MSSA clones, leading to the incidence of CA-MRSA [11]. Thus, determination of the genetic characteristics of MSSA strains is crucial for further understanding of the epidemiology of MSSA and even of CA-MRSA.

S. aureus is the most common cause of bacterial keratitis (corneal infection) and conjunctivitis (conjunctival infection) [12,13]. Although the Antibiotic Resistance Monitoring in Ocular Microorganisms (ARMOR) study reported that 65.1% of *S. aureus* isolates were MSSA [14], studies on MSSA ocular infections have generally been limited to small case series. Previously, we conducted a 10-year retrospective study on *S. aureus* ocular infections [15] and observed that the percentages of ocular infections caused by MSSA and by MRSA were approximately equal; both strains caused a similar spectrum of diseases, although MRSA was more frequently associated with eyelid infections (16.7% vs. 24.5%, $p = 0.040$). In our subsequent research, we focused solely on MRSA ocular infections [16] and performed molecular typing on MRSA isolates [17]. Because the importance of MSSA should not be overlooked, in this study, we investigated the clinical features, molecular characteristics, and antibiograms of MSSA ocular infections and compared them between CA-MSSA and HA-MSSA isolates.

2. Results

2.1. Clinical Characteristics of MSSA Ocular Infections

During this 8-year study period, MSSA strains were isolated from 104 patients. Of the 104 patients with MSSA, 46 (44.2%) were classified as having CA-MSSA, and 58 (55.8%) were classified as having HA-MSSA. Table 1 presents a summary of the demographic characteristics and clinical features of the patients with MSSA ocular infections. The patients with HA-MSSA were significantly more likely to have underlying conditions, such as malignancy or current infection, compared with those with CA-MSSA ($p = 0.006$ and <0.001, respectively). Regarding local risk factors, a significantly higher proportion of the patients with HA-MSSA had a history of ocular surgery, and a significantly higher proportion of the patients with CA-MSSA used contact lenses. Keratitis was the most common ocular disease caused by CA-MSSA, followed by conjunctivitis; however, this order was reversed in the HA-MSSA group (Table 2). The rates of keratitis and conjunctivitis caused by the CA-MSSA and HA-MSSA strains were significantly different (keratitis: 63.0% vs. 29.3%, $p < 0.001$; conjunctivitis: 10.9% vs. 39.7%, $p = 0.001$). Most of the patients with CA-MSSA received outpatient or emergency department treatment (69.6%), whereas most of the patients with HA-MSSA infection received inpatient treatment (56.9%).

Table 1. Comparison of demographics and systemic and local factors between patients infected with community-associated and health-care-associated methicillin-susceptible *Staphylococcus aureus* ocular isolates.

	CA (*n* = 46) No. (%)	HA (*n* = 58) No. (%)	*p* Value
Demographics			
Age in years, mean ± SD (range)	44.6 ± 24.0 (0.1–84)	52.6 ± 25.0 (0.1–95)	0.102
Sex (male/female)	22/24 (47.8/52.2)	25/33 (43.1/56.9)	0.694
Underlying Condition			
Diabetes mellitus	9 (19.6)	15 (25.9)	0.491
Hypertension	13 (28.3)	20 (34.5)	0.532
Pulmonary disease	3 (6.5)	6 (10.3)	0.728
Renal disease	4 (8.7)	6 (10.3)	1
Liver disease	2 (4.3)	5 (8.6)	0.46

Table 1. Cont.

	CA (n = 46) No. (%)	HA (n = 58) No. (%)	p Value
Malignancy	1 (2.2)	12 (20.7)	**0.006**
Immunodeficiency	1 (2.2)	5 (8.6)	0.224
Current infection [a]	0 (0)	19 (32.9)	**<0.001**
Recent antibiotic use	1 (2.2)	15 (25.9)	**0.001**
Alcoholic	2 (4.3)	3 (5.2)	1
Ocular history			
Contact lens use	8 (17.4)	1 (1.7)	**0.010**
Ocular surface disease	11 (23.9)	23 (39.7)	0.098
Surgery	9 (19.6)	37 (63.8)	**<0.001**
Trauma	8 (17.4)	9 (15.5)	0.797

[a] Nonocular infection; CA = community-associated; HA = health-care-associated.

Table 2. Comparison of diagnoses and treatments between community-associated and health-care-associated methicillin-susceptible *Staphylococcus aureus* ocular isolates.

	CA (n = 46) No. (%)	HA (n = 58) No. (%)	p Value
Diagnosis			
Lid disorder	4 (8.7)	4 (6.9)	0.730
Conjunctivitis	5 (10.9)	23 (39.7)	**0.001**
Keratitis	29 (63.0)	17 (29.3)	**<0.001**
Endophthalmitis	3 (6.5)	2 (3.4)	0.653
Wound infection)	1 (2.2)	5 (8.6)	0.224
Lacrimal system disorder	4 (8.7)	4 (6.9)	0.730
Others (%)	0 (0)	3 (5.2)	0.253
Treatment			
Surgical intervention	5 (10.9)	11 (19)	0.288
Inpatient	14 (30.4)	33 (56.9)	**0.009**
Outpatient/ED	32 (69.6)	25 (43.1)	**0.010**

CA = community-associated; ED = emergency department; HA = health-care-associated.

2.2. Molecular Typing

Table 3 presents the molecular typing of the MSSA isolates. We identified a total of 14 pulsotypes (Figure 1), of which BA was the most common (33.7%), followed by F (19.2%). No specific pulsotype distinguished HA isolates from CA isolates. Multilocus sequence typing (MLST) was performed on 39 isolates selected from each pulsed-field gel electrophoresis (PFGE) type and from four untypeable samples. We identified 21 sequence types (STs); a phylogenic tree of these types is presented in Figure 2. For pulsotype F, all six selected isolates belonged to ST15. For pulsotype BA, cluster complex (CC) 7, including ST7 and ST6427, was the dominant MLST type (5/7). Seven isolates (6.7%), comprising five CA and two HA isolates, contained Panton-Valentine leukocidin (PVL) genes. The PVL-positive isolates contained ST1232 (n = 2), ST59, ST338, ST6426, and ST672, as well as one untypeable isolate (Table 3).

Table 3. Molecular characteristics of 104 methicillin-susceptible *Staphylococcus aureus* ocular isolates stratified by pulsotype.

Pulsotypes	BA	F	AX	BW	D	Others
No. isolates (n = 104)	35 (33.7%)	20 (19.2%)	8 (7.7%)	8 (7.7%)	7 (6.7%)	26 (25%)
CA (n = 46)	13 (28.3%)	10 (21.7%)	4 (8.7%)	3 (6.5%)	5 (10.9%)	11 (23.9%)
HA (n = 58)	22 (37.9%)	10 (17.2%)	4 (6.9%)	5 (8.6%)	2 (3.4%)	15 (25.9%)
p-value	0.404	0.621	0.730	1	0.237	1
PVL-positive (n = 7)	0	0	1	1	2	3

Table 3. Cont.

Pulsotypes	BA	F	AX	BW	D	Others
Sequence type	1 (1/7), 7 (4/7), 6427 (1/7), 6457 (1/7)	15 (6/6)	188 (2/3), 6426 [a] (1/3)	Untypeable [a] (4/4)	59 [a] (2/4), 97 (1/3), 338 [a] (1/4)	1232 [aa] (3), 59, 1281, 72, 30 (2), 96, 398, 508, 509, 573, 672 [a], 6453, untypeable

[a]: Sequence type of PVL-positive isolates; [aa]: Two isolates with PVL were ST 1232; CA = community-associated; HA = health-care-associated; PVL = Panton-Valentine leukocidin genes.

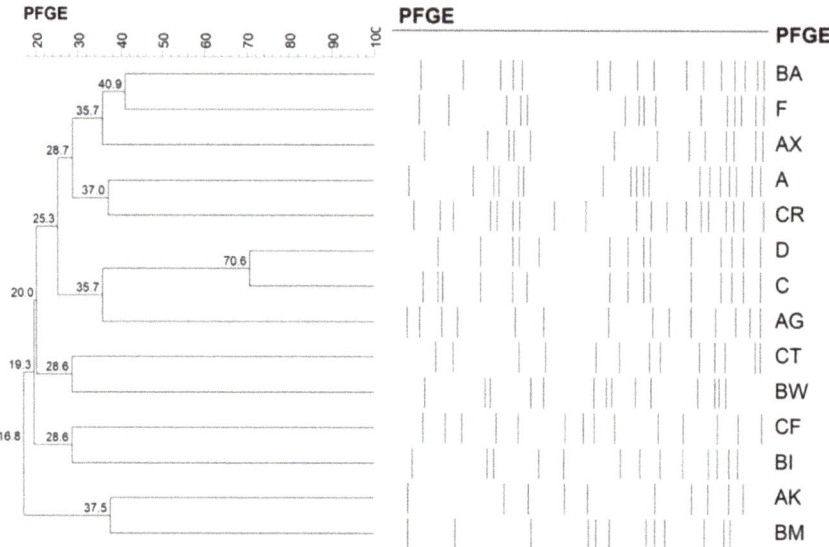

Figure 1. Dendrogram of pulsed-field gel electrophoresis cluster analysis of 104 methicillin-susceptible *Staphylococcus aureus* ocular isolates, classified into 14 pulsotypes.

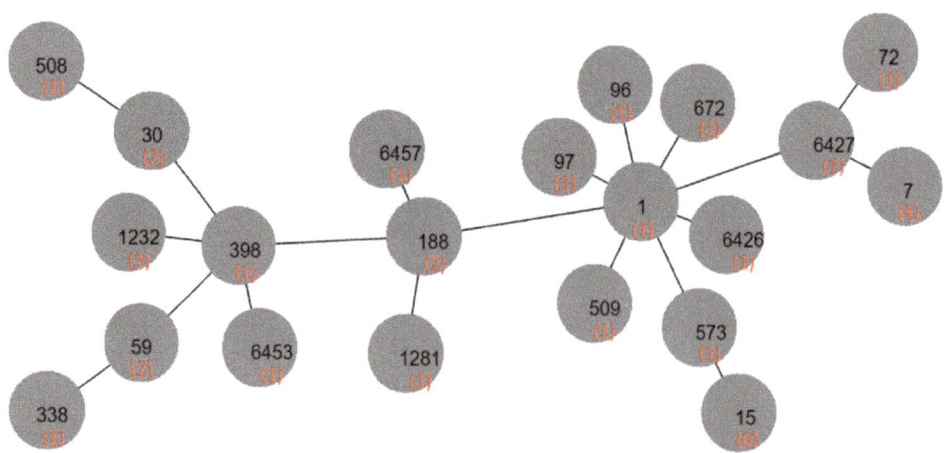

Figure 2. Phylogenic tree of multilocus sequence types of methicillin-susceptible *Staphylococcus aureus* ocular isolates. Bracketed numbers represent the number of isolates.

2.3. Drug Susceptibility Test

All the CA-MSSA and HA-MSSA isolates were susceptible to linezolid, tigecycline, vancomycin, and teicoplanin. In addition, all the CA-MSSA isolates and more than 95% of the HA-MSSA isolates were susceptible to sulfamethoxazole–trimethoprim (TMP–SMX) and fluoroquinolones. The CA-MSSA and HA-MSSA strains exhibited similar antibiotic susceptibility levels, except for susceptibility to erythromycin (63% vs. 81%, $p = 0.047$) (Figure 3).

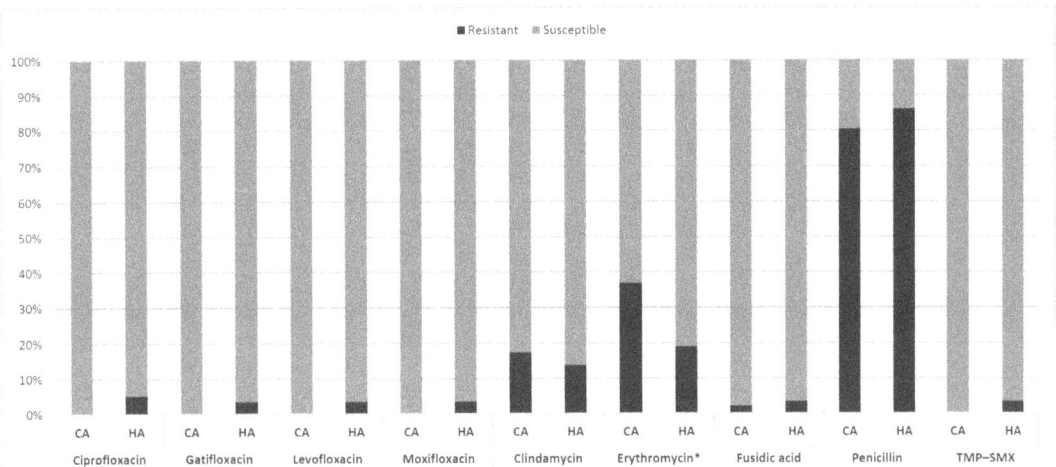

Figure 3. Antibiotic susceptibility of community-associated and health-care-associated methicillin-susceptible *Staphylococcus aureus* ocular isolates. CA = community-associated; HA = health-care-associated; TMP–SMX = sulfamethoxazole–trimethoprim; * $p = 0.047$.

3. Discussion

Few studies have explored the clinical and molecular characteristics of MSSA infections, particularly ocular infections. In this study, we investigated the clinical and microbiological characteristics of MSSA ocular infections in Taiwan and compared them between CA-MSSA and HA-MSSA ocular isolates, which have rarely been compared in previous studies. We observed that over half of MSSA ocular infections were HA. Keratitis and conjunctivitis accounted for most of the clinical manifestations of the MSSA ocular infections; significantly more patients with CA infections presented with keratitis, whereas those with HA infections more often presented with conjunctivitis. The molecular characteristics of the MSSA isolates were relatively diverse, and no specific genotype differentiated CA-MSSA from HA-MSSA. The antibiograms of the CA-MSSA and HA-MSSA isolates were similar.

In this study, a higher rate of HA-MSSA than of CA-MSSA was observed in the MSSA ocular isolates (55.8%), which is consistent with the findings of our previous 10-year study on *S. aureus* ocular infections [15] but contradicts the findings of a study on pediatric MSSA infections in our hospital in 2015, in which the rate of CA-MSSA was reported to be 71.8% [7]. As expected, patients with HA-MSSA ocular infections exhibited a higher percentage of comorbidities such as malignancy and recent non-ophthalmic infections. Regarding local factors, we observed that a significantly higher proportion of the patients with HA-MSSA infections had undergone ocular surgery; however, a significantly higher proportion of the patients with CA-MSSA infections had a history of contact lens use.

S. aureus is the leading cause of bacterial keratitis and is the most common pathogen isolated from patients with conjunctivitis [12,13]. Marangon and Miller studied 1230 MSSA isolates from corneal and conjunctival infections between 1990 and 2001 and observed that the rate of keratitis was 62–65% [18]. Our previous 10-year study [15] and the current study have also confirmed that keratitis and conjunctivitis are common in MSSA ocular infections. In the present study, the patients with CA-MSSA infections exhibited a significantly higher rate of keratitis than did those with HA-MSSA infections, but HA-MSSA isolates caused higher rates of conjunctivitis (followed by keratitis). However, these results are different from those of our previous 10-year report on MRSA ocular infections, in which HA-MRSA isolates caused more keratitis but less conjunctivitis than did CA-MRSA isolates [16].

Regarding molecular characteristics, the MSSA isolates in this study were polyclonal, and none of the clones exhibited significant differences between the CA and HA isolates. Our study revealed that ST7/pulsotype BA, ST15/pulsotype F, and ST188/pulsotype AX accounted for half of the isolates, which is consistent with the findings of previous studies on MSSA nonocular infections in Taiwan [7,10,19]. However, among the samples employed in the previous studies, ST188 was the most common sequence type, whereas in the present study, ST7/pulsotype BA was the most frequent sequence type, which is consistent with the findings of a similar study from China [20]. Although Chen et al. [7] reported that ST15/pulsotype F isolates were more frequently observed in CA-MSSA than in HA-MSSA ($p = 0.064$) among pediatric patients, other research on adult patients [10] did not reveal differences between the genotypes of CA-MSSA and HA-MSSA isolates in Taiwan. Hesje et al. [21] and Peterson et al. [22] have investigated the molecular characteristics of *S. aureus* ocular isolates in the US, and both studies have demonstrated the polyclonality of MSSA ocular isolates. However, Peterson et al. reported that two major USA clonal complexes, CC5 and CC8, were the most frequently detected clones among MSSA and MRSA keratitis isolates. This discrepancy in the molecular characteristics of MSSA ocular isolates indicates geographic variation.

PVL, a bicomponent pore-forming toxin targeting phagocytic leukocytes, is regarded as a marker of CA-MRSA. In this study, we identified PVL genes in seven MSSA isolates (6.7%). The low prevalence of PVL-positive MSSA strains is comparable to that observed in a global clinical trial [8], which reported a rate of 8.8% (118/1334) in MSSA isolates from 2004 to 2005, and to that in a study conducted in central Taiwan, which reported a rate of 5.4% [19]. The endemic CA-MRSA clone might have originated from MSSA. The previous study conducted in Taiwan demonstrated that PVL-positive ST59 MSSA shared a similar genetic profile with PVL-positive ST59 MRSA [23], the endemic CA-MRSA clone. Chen et al. [7] also reported that among pediatric patients at our hospital, 9 of 11 PVL-positive MSSA isolates belonged to ST59. However, the STs of PVL-positive MSSA isolates were heterogenous in this study, with only two belonging to CC59 (ST59 and ST338). Further research is warranted to determine whether the PVL-positive MSSA isolates derived from ocular and nonocular infections are different.

In the present study, the HA-MSSA and CA-MSSA isolates both exhibited high susceptibility to most of the tested antibiotics except for clindamycin and erythromycin. The antibiograms of the MSSA ocular isolates were determined to parallel those of nonocular MSSA isolates in Taiwan. We tested the isolates' antibiotic susceptibilities to fluoroquinolones, the first-line treatment of ocular infections, and observed that the susceptibility rates to fluoroquinolones all exceeded 90%. This rate is higher than that (approximately 80%) reported by Asbell et al. in their study on MSSA infection in the ocular Tracking Resistance in U.S. Today (ocular TRUST) program [24]. Taiwan's National Health Insurance Administration strictly regulates the use of ophthalmic fluoroquinolones solutions. Such solutions are reserved for the treatment of severe bacterial infections such as corneal ulcers but are not permitted for prophylactic purposes or the treatment of mild infections, which might have contributed to the relatively high susceptibility of the isolates to fluoroquinolones in our isolates. Although the increasing resistance of MSSA ocular isolates to fluoroquinolones reported by Marangon et al. [18] is a concern, fluoroquinolones might remain the treatment of choice for MSSA ocular infections, at least in Taiwan.

The emergence of drug resistance necessitates innovative therapeutic approaches. Recently, new ophthalmic solutions have been developed. Povidone–iodine 0.6% was determined to exhibit higher bactericidal activity than does povidone–iodine 5% [25]; a commercial solution containing povidone–iodine 0.6% (IODIM, Medivis, Catania, Italy) exhibited high antimicrobial activity against *S. aureus*, *S. epidermidis*, and *Pseudomonas aeruginosa* in vitro [26]. Another solution containing hexamidine diisethionate 0.05% (Keratosept, Bruschettini, Genoa, Italy) also exhibited rapid antibacterial activity against multiresistant *S. aureus*, *S. epidermidis*, and *Candida* species [27]. In addition to bactericidal agents, antibiotic combinations represent an alternative method for treating multiresistant pathogens. Nasir et al. [28] reported that levofloxacin combined with ceftazidime was successful against MRSA isolates. In future research, these treatments should be studied in vivo to determine if they are as effective as they are in vitro.

In contrast to the differentiation of MRSA isolates, differentiating between CA-MSSA and HA-MSSA ocular isolates is difficult. In our previous study on MRSA ocular infections, we found that patients with CA-MRSA ocular infections were, on average, younger and more frequently diagnosed with eyelid disorders than those with HA-MRSA infections [16]. The dominant clones for CA-MRSA and HA-MRSA isolates were ST59 and ST239, respectively [17]; both CA-MRSA and HA-MRSA were multiresistant, but TMP–SMX exhibited high activity against CA-MRSA [16,17]. These results are consistent with those of studies on MRSA nonocular infections in Taiwan. The epidemiological characteristics of each isolate could distinguish CA-MRSA from HA-MRSA isolates and be used to evaluate each isolate's clinical or pathogenic implications. However, the present study demonstrated that the patients with CA-MSSA and HA-MSSA ocular infections exhibited similar demographic characteristics, microbiological characteristics, and clinical features, except for the spectrum of diseases and admission rates. This implies that the distinction between CA-MSSA and HA-MSSA is blurred, partially due to the transmission of CA strains into HA facilities. Chen et al. [7] investigated the clinical features and molecular characteristics of pediatric MSSA infections in our hospital and determined that more patients with CA-MSSA presented with skin and soft tissue infections than did those with HA-MSSA infections, a clinical feature that is like those of pediatric MRSA infections. The molecular characteristics of the MSSA isolates derived from the pediatric patients were diverse, but one clone (ST15/pulsotype F) exhibited a borderline significant difference between the CA-MSSA and HA-MSSA isolates. Furthermore, the CA-MSSA isolates exhibited a significantly higher susceptibility rate to TMP–SMX (100%) than did the HA-MSSA isolates (95%). Further research involving larger sample sizes is necessary to determine whether the discrepancy between the results of the present study and those of the pediatric MSSA study [7] conducted in the same hospital is due to tissue tropism or other causes.

Our study has several limitations. First, although the isolates were prospectively collected, the clinical data were retrospectively reviewed. Therefore, the patients might have been misclassified due to incomplete evaluation of risk factors. Nevertheless, we were still able to observe some significant differences between the patients infected with the CA and HA-MSSA isolates. Second, the sample size might not have been sufficiently large, but the correlation between pulsotypes and STs is similar to that recorded in the study on pediatric MSSA infection in Taiwan [7]. Third, determining whether the isolate was a contaminant or pathogen was difficult because *S. aureus* is a common colonizer of the ocular surface. However, all the isolates included in this study were clinical samples collected from patients with active ocular infections. Finally, this study was conducted in a single tertiary-care hospital; therefore, the results might not be generalizable to other populations. However, because our hospital is the largest referral hospital in Taiwan, the results of this study may still generally reflect the epidemiology of MSSA infection. Additional prospective studies involving patients from more hospitals are warranted.

4. Materials and Methods

4.1. Ethics

The study adhered to the tenets of the Declaration of Helsinki and was approved by the Institutional Review Board (IRB) of Chang Gung Memorial Hospital (CGMH), a tertiary medical center in Taoyuan, Taiwan (IRB 107-2346C). The requirement for written informed consent was waived due to the anonymous analysis of the data.

4.2. Study Population and Data Collection

From 1 January 2010 to 31 December 2017, clinical *S. aureus* isolates were prospectively collected from patients with ocular infections and stored in the microbiology laboratory of CGMH. The medical records of the patients were retrospectively reviewed for demographic and clinical information. If more than one isolate was collected from a single patient, only the first was included for analysis. The MSSA infections were classified as HA or CA based on the definitions employed by the CDC since 2000 [29]. The clinical criteria for HA-MSSA consist of an infection identified 48 h after hospitalization; a history of healthcare exposure within 1 year prior to the presentation including admission, surgery, dialysis, or residency in a long-term care facility; and the use of permanent indwelling catheters or percutaneous devices. CA-MSSA infections are defined as MSSA infections identified within 48 h of hospitalization in patients without a history of healthcare exposure within 1 year prior to the presentation.

The patients' underlying conditions and ocular histories were recorded. The underlying conditions that we screened for were diabetes mellitus, hypertension, pulmonary disease, renal disease, liver disease, malignancy, immunodeficiency, current infection, recent antibiotic use, and alcoholism; we also screened for recent antibiotic use. The ocular histories included contact lens use, ocular surface disease, surgery, and trauma.

Based on the ocular structures involved, the manifestations were categorized into seven diagnoses: lid disorder, conjunctivitis, keratitis, endophthalmitis, wound infection, lacrimal system disorder, and others (including blebitis, buckle or implant infection, and scleral ulcer). If a patient was diagnosed with more than one ocular infection, the primary pathology or the most severe diagnosis was chosen.

4.3. Molecular Characteristics

Molecular analysis methods employed in this study included PFGE by SmaI digestion, PVL gene detection, and MLST, with approaches consistent with those described in detail by Lina et al. [30] and Enright et al. [31]. All the MSSA isolates were molecularly characterized on the basis of PFGE and PVL genes. MLST was performed on selected isolates with representative PFGE patterns. The ST was determined according to each isolate's allelic profile.

4.4. Drug Susceptibility Test

We evaluated the antimicrobial susceptibility of the MSSA isolates to antibiotics (namely penicillin, oxacillin, TMP–SMX, clindamycin, erythromycin, fusidic acid, teicoplanin, vancomycin, tigecycline, and linezolid) through the disk diffusion method according to the Clinical and Laboratory Standards Institute's standards for antimicrobial susceptibility testing [32]. An E test (bioMerieux, Marcy-I'Etoile, France) was also used to determine the isolates' susceptibility to fluoroquinolones, namely ciprofloxacin, levofloxacin, gatifloxacin, and moxifloxacin. Oxacillin was used instead of methicillin to test for β-lactam resistance.

4.5. Statistical Analysis

All statistical analyses were performed using SPSS (Version 19.0; IBM, Armonk, NY, USA). The nominal variables were analyzed through a chi-square test, and the continuous variables were analyzed using Student's t test. The variables are presented either as a mean ± standard deviation or as a percentage. The correlation between clinical presentations and the classification of MSSA infections as HA or CA was measured using Fisher's exact test. A two-tailed p value of less than 0.05 was considered statistically significant.

5. Conclusions

More than half of the MSSA ocular infections included in this study were classified as HA-MSSA. CA-MSSA caused a higher rate of keratitis than did HA-MSSA, whereas HA-MSSA caused a higher rate of conjunctivitis. Although the molecular characteristics of the MSSA isolates indicated that the isolates were genetically diverse, ST7, ST15, and ST188 were frequently observed in the MSSA ocular infections. Both CA- and HA-MSSA strains exhibited high susceptibility to fluoroquinolones in Taiwan. Physicians should be familiar with the epidemiology, spectrum of diseases, and antibiotic susceptibility patterns of MSSA ocular infections in their local areas. Although we could not clearly differentiate HA-MSSA from CA-MSSA in this study, the results provide information that may be used to enhance local public health policy as well as knowledge on epidemic MSSA clones worldwide. Further research should include larger sample sizes and involve more hospitals to deepen the understanding of the molecular characteristics and clinical features of MSSA ocular infections.

Author Contributions: Conceptualization, Y.-C.H. and C.-H.H.; methodology, Y.-C.H. and C.-H.H.; formal analysis, Y.-L.C. and E.Y.-C.K.; investigation, Y.-L.C. and E.Y.-C.K.; data curation, L.-K.Y., D.H.K.M., H.-Y.T., H.-C.C. and K.-H.H.; writing—original draft preparation, Y.-L.C.; writing—review and editing, Y.-L.C. and C.-H.H.; supervision, Y.-C.H.; project administration, C.-H.H.; funding acquisition, C.-H.H. All authors have read and agreed to the published version of the manuscript.

Funding: The study was supported by Chang Gung Memorial Hospital, Taiwan (CMRPG1I0021).

Institutional Review Board Statement: The study was conducted according to the guidelines of the Declaration of Helsinki, and approved by the Institutional Review Board of Chang Gung Memorial Hospital (CGMH) (IRB 107-2346C).

Informed Consent Statement: Patient consent was waived due to the anonymous analysis of the data.

Data Availability Statement: The data presented in this study are available upon request.

Acknowledgments: In this section, you can acknowledge any support given which is not covered by the author contribution or funding sections. This may include administrative and technical support, or donations in kind (e.g., materials used for experiments).

Conflicts of Interest: The authors declare no conflict of interest.

References

1. Magill, S.S.; O'Leary, E.; Janelle, S.J.; Thompson, D.L.; Dumyati, G.; Nadle, J.; Wilson, L.E.; Kainer, M.A.; Lynfield, R.; Greissman, S.; et al. Changes in Prevalence of Health Care–Associated Infections in U.S. Hospitals. *N. Eng. J. Med.* **2018**, *379*, 1732–1744. [CrossRef]
2. Turner, N.A.; Sharma-Kuinkel, B.K.; Maskarinec, S.A.; Eichenberger, E.M.; Shah, P.P.; Carugati, M.; Holland, T.L.; Fowler, V.G. Methicillin-resistant Staphylococcus aureus: An overview of basic and clinical research. *Nat. Rev. Microbiol.* **2019**, *17*, 203–218. [CrossRef]
3. Wong, J.W.; Ip, M.; Tang, A.; Wei, V.W.; Wong, S.Y.; Riley, S.; Read, J.M.; Kwok, K.O. Prevalence and risk factors of community-associated methicillin-resistant *Staphylococcus aureus* carriage in Asia-Pacific region from 2000 to 2016: A systematic review and meta-analysis. *Clin. Epidemiol.* **2018**, *10*, 1489–1501. [CrossRef] [PubMed]
4. King, J.M.; Kulhankova, K.; Stach, C.S.; Vu, B.G.; Salgado-Pabón, W. Phenotypes and Virulence among *Staphylococcus aureus* USA100, USA200, USA300, USA400, and USA600 Clonal Lineages. *Msphere* **2016**, *1*, e00071-16. [CrossRef]
5. Jackson, K.A.; Gokhale, R.H.; Nadle, J.; Ray, S.M.; Dumyati, G.; Schaffner, W.; Ham, D.C.; Magill, S.S.; Lynfield, R.; See, I. Public Health Importance of Invasive Methicillin-sensitive *Staphylococcus aureus* Infections: Surveillance in 8 US Counties, 2016. *Clin. Infect. Dis.* **2020**, *70*, 1021–1028. [CrossRef] [PubMed]
6. Kourtis, A.P.; Hatfield, K.; Baggs, J.; Mu, Y.; See, I.; Epson, E.; Nadle, J.; Kainer, M.A.; Dumyati, G.; Petit, S.; et al. Vital Signs: Epidemiology and Recent Trends in Methicillin-Resistant and in Methicillin-Susceptible *Staphylococcus aureus* Bloodstream Infections—United States. *MMWR Morb. Mortal. Wkly. Rep.* **2019**, *68*, 214–219. [CrossRef]
7. Chen, Y.J.; Chen, P.A.; Chen, C.J.; Huang, Y.C. Molecular characteristics and clinical features of pediatric methicillin-susceptible *Staphylococcus aureus* infection in a medical center in northern Taiwan. *BMC Infect. Dis.* **2019**, *19*, 402. [CrossRef]

8. Goering, R.V.; Shawar, R.M.; Scangarella, N.E.; O'Hara, F.P.; Amrine-Madsen, H.; West, J.M.; Dalessandro, M.; Becker, J.A.; Walsh, S.L.; Miller, L.A.; et al. Molecular epidemiology of methicillin-resistant and methicillin-susceptible *Staphylococcus aureus* isolates from global clinical trials. *J. Clin. Microbiol.* **2008**, *46*, 2842–2847. [CrossRef]
9. Chen, F.-J.; Siu, L.-K.K.; Lin, J.-C.; Wang, C.-H.; Lu, P.-L. Molecular typing and characterization of nasal carriage and community-onset infection methicillin-susceptible *Staphylococcus aureus* isolates in two Taiwan medical centers. *BMC Infect. Dis.* **2012**, *12*, 343. [CrossRef]
10. Chen, P.-Y.; Chuang, Y.-C.; Wang, J.-T.; Chang, S.-C. Impact of Prior Healthcare-Associated Exposure on Clinical and Molecular Characterization of Methicillin-Susceptible *Staphylococcus aureus* Bacteremia. *Medicine* **2015**, *94*, e474. [CrossRef] [PubMed]
11. De Kraker, M.E.; Jarlier, V.; Monen, J.C.; Heuer, O.E.; van de Sande, N.; Grundmann, H. The changing epidemiology of bacteraemias in Europe: Trends from the European Antimicrobial Resistance Surveillance System. *Clin. Microbiol. Infect.* **2013**, *19*, 860–868. [CrossRef]
12. Gudmundsson, O.G.; Ormerod, L.D.; Kenyon, K.R.; Glynn, R.J.; Baker, A.S.; Haaf, J.; Lubars, S.; Abelson, M.B.; Boruchoff, S.A.; Foster, C.S.; et al. Factors influencing predilection and outcome in bacterial keratitis. *Cornea* **1989**, *8*, 115–121. [CrossRef]
13. Knauf, H.P.; Silvany, R.; Southern, P.M., Jr.; Risser, R.C.; Wilson, S.E. Susceptibility of corneal and conjunctival pathogens to ciprofloxacin. *Cornea* **1996**, *15*, 66–71. [CrossRef] [PubMed]
14. Asbell, P.A.; Sanfilippo, C.M.; Sahm, D.F.; DeCory, H.H. Trends in Antibiotic Resistance Among Ocular Microorganisms in the United States From 2009 to 2018. *JAMA Ophtalmol.* **2020**, *138*, 439–450. [CrossRef] [PubMed]
15. Chuang, C.C.; Hsiao, C.H.; Tan, H.Y.; Ma, D.H.; Lin, K.K.; Chang, C.J.; Huang, Y.C. *Staphylococcus aureus* ocular infection: Methicillin-resistance, clinical features, and antibiotic susceptibilities. *PLoS ONE* **2012**, *8*, e42437. [CrossRef] [PubMed]
16. Hsiao, C.H.; Chuang, C.C.; Tan, H.Y.; Ma, D.H.; Lin, K.K.; Chang, C.J.; Huang, Y.C. Methicillin-resistant *Staphylococcus aureus* ocular infection: A 10-year hospital-based study. *Ophtalmology* **2012**, *119*, 522–527. [CrossRef]
17. Kang, Y.C.; Hsiao, C.H.; Yeh, L.K.; Ma, D.H.K.; Chen, P.Y.F.; Lin, H.C.; Tan, H.Y.; Chen, H.C.; Chen, S.Y.; Huang, Y.C. Methicillin-Resistant *Staphylococcus aureus* Ocular Infection in Taiwan: Clinical Features, Genotyping, and Antibiotic Susceptibility. *Medicine* **2015**, *94*, e1620. [CrossRef]
18. Marangon, F.B.; Miller, D.; Muallem, M.S.; Romano, A.C.; Alfonso, E.C. Ciprofloxacin and levofloxacin resistance among methicillin-sensitive *Staphylococcus aureus* isolates from keratitis and conjunctivitis. *Am. J. Ophtalmol.* **2004**, *137*, 453–458. [CrossRef]
19. Ho, C.-M.; Lin, C.-Y.; Ho, M.-W.; Lin, H.-C.; Peng, C.-T.; Lu, J.-J. Concomitant genotyping revealed diverse spreading between methicillin-resistant *Staphylococcus aureus* and methicillin-susceptible *Staphylococcus aureus* in central Taiwan. *J. Microbiol. Immunol. Infect.* **2016**, *49*, 363–370. [CrossRef]
20. Yu, F.; Li, T.; Huang, X.; Xie, J.; Xu, Y.; Tu, J.; Qin, Z.; Parsons, C.; Wang, J.; Hu, L.; et al. Virulence gene profiling and molecular characterization of hospital-acquired *Staphylococcus aureus* isolates associated with bloodstream infection. *Diagn. Microbiol. Infect. Dis.* **2012**, *74*, 363–368. [CrossRef]
21. Hesje, C.K.; Sanfilippo, C.M.; Haas, W.; Morris, T.W. Molecular epidemiology of methicillin-resistant and methicillin-susceptible *Staphylococcus aureus* isolated from the eye. *Curr. Eye Res.* **2011**, *36*, 94–102. [CrossRef] [PubMed]
22. Peterson, J.C.; Durkee, H.; Miller, D.; Maestre-Mesa, J.; Arboleda, A.; Aguilar, M.C.; Relhan, N.; Flynn, H.W., Jr.; Amescua, G.; Parel, J.M.; et al. Molecular epidemiology and resistance profiles among healthcare- and community-associated *Staphylococcus aureus* keratitis isolates. *Infect. Drug Resist* **2019**, *12*, 831–843. [CrossRef] [PubMed]
23. Chen, F.-J.; Hiramatsu, K.; Huang, I.W.; Wang, C.-H.; Lauderdale, T.-L.Y. Panton-Valentine leukocidin (PVL)-positive methicillin-susceptible and resistant *Staphylococcus aureus* in Taiwan: Identification of oxacillin-susceptible mecA-positive methicillin-resistant. *S. Aureus. Diagn. Microbiol. Infect. Dis.* **2009**, *65*, 351–357. [CrossRef] [PubMed]
24. Asbell, P.A.; Colby, K.A.; Deng, S.; McDonnell, P.; Meisler, D.M.; Raizman, M.B.; Sheppard, J.D., Jr.; Sahm, D.F. Ocular TRUST: Nationwide antimicrobial susceptibility patterns in ocular isolates. *Am. J. Ophtalmol.* **2008**, *145*, 951–958. [CrossRef] [PubMed]
25. Musumeci, R.; Bandello, F.; Martinelli, M.; Calaresu, E.; Cocuzza, C.E. In vitro bactericidal activity of 0.6% povidone-iodine eye drops formulation. *Eur. J. Ophtalmol.* **2019**, *29*, 673–677. [CrossRef] [PubMed]
26. Pinna, A.; Donadu, M.G.; Usai, D.; Dore, S.; D'Amico-Ricci, G.; Boscia, F.; Zanetti, S. In vitro antimicrobial activity of a new ophthalmic solution containing povidone-iodine 0.6% (IODIM®). *Acta Ophtalmol.* **2020**, *98*, e178–e180. [CrossRef]
27. Pinna, A.; Donadu, M.G.; Usai, D.; Dore, S.; Boscia, F.; Zanetti, S. In Vitro Antimicrobial Activity of a New Ophthalmic Solution Containing Hexamidine Diisethionate 0.05% (Keratosept). *Cornea* **2020**, *39*, 1415–1418. [CrossRef]
28. Nasir, S.; Vohra, M.S.; Gul, D.; Swaiba, U.E.; Aleem, M.; Mehmood, K.; Andleeb, S. Novel Antibiotic Combinations of Diverse Subclasses for Effective Suppression of Extensively Drug-Resistant Methicillin-Resistant *Staphylococcus aureus* (MRSA). *Int. J. Microbiol.* **2020**, *2020*, 8831322. [CrossRef]
29. Morrison, M.A.; Hageman, J.C.; Klevens, R.M. Case definition for community-associated methicillin-resistant Staphylococcus aureus. *J. Hosp. Infect.* **2006**, *62*, 241. [CrossRef]
30. Lina, G.; Piémont, Y.; Godail-Gamot, F.; Bes, M.; Peter, M.O.; Gauduchon, V.; Vandenesch, F.; Etienne, J. Involvement of Panton-Valentine leukocidin-producing *Staphylococcus aureus* in primary skin infections and pneumonia. *Clin. Infect. Dis.* **1999**, *29*, 1128–1132. [CrossRef]
31. Enright, M.C.; Day, N.P.; Davies, C.E.; Peacock, S.J.; Spratt, B.G. Multilocus sequence typing for characterization of methicillin-resistant and methicillin-susceptible clones of Staphylococcus aureus. *J. Clin. Microbiol.* **2000**, *38*, 1008–1015. [CrossRef] [PubMed]
32. Clinical and Laboratory Standards Institute. *Performance Standards for Antimicrobial Susceptibility Testing*; Twenty-Third Informnational Supplement; Clinical and Laboratory Standards Institute: Wayne, PA, USA, 2013.

Article

The Impact of Antibiotic Usage Guidelines, Developed and Disseminated through Internet, on the Knowledge, Attitude and Prescribing Habits of Orthokeratology Contact Lens Practitioners in China

Zhi Chen [1,2,3,*,†], Jifang Wang [1,2,3,†], Jun Jiang [4], Bi Yang [5] and Pauline Cho [6]

1. Eye Institute and Department of Ophthalmology, Eye and ENT Hospital, Fudan University, Shanghai 200031, China; carol1979@yeah.net
2. NHC Key Laboratory of Myopia, Fudan University, Shanghai 200031, China
3. Key Laboratory of Myopia, Chinese Academy of Medical Sciences, Shanghai 200031, China
4. Department of Contact Lens Clinic, The Eye Hospital of Wenzhou Medical University, Wenzhou 325027, China; jjhsj@hotmail.com
5. Department of Optometry and Vision Sciences, West China School of Medicine, Sichuan University, Chengdu 610017, China; yangbi19830418@126.com
6. School of Optometry, The Hong Kong Polytechnic University, Hong Kong SAR, China; pauline.cho@connect.polyu.hk
* Correspondence: peter459@aliyun.com
† Z.C. and J.W. contributed equally to this study and were considered co-first authors.

Abstract: It has been previously reported that the improper prescribing of antibiotic eye drops is common among orthokeratology (ortho-k) practitioners. Guidelines have since been developed and disseminated to improve their understanding and implementation of antibiotic prescriptions. This study aimed to investigate the influence of these guidelines on the knowledge, attitude, and prescribing habits of ortho-k practitioners by means of a questionnaire, which was administered nationwide via an official online account to eye care practitioners (ECPs) involved in ortho-k lens fitting, 548 of whom completed the survey. Differences in characteristics before and after the dissemination of the guidelines and between the groups were explored using χ^2 tests. The relationship between prescribing habits and demographics was analyzed using stepwise logistic regression models. The implementation of the guidelines significantly improved the overall prescribing habits of ECPs ($p < 0.001$), especially for prophylactic antibiotic use before and after ortho-k lens wear ($p < 0.001$). Most ECPs who prescribed antibiotics properly displayed significantly better knowledge of correct antibiotic use, which in turn affected the compliance in their ortho-k patients ($p < 0.001$). The ECPs' occupations (professionals other than ophthalmologists and optometrists, including nurses and opticians), clinical setting (distributor fitting centers), and age (younger than 25 years) were risk factors for the misuse of antibiotics. Although the implementation of the antibiotic guidelines significantly improved overall prescribing habits, some practitioners' prescribing behavior still needs improvement. A limitation of this study was that all questions were mandatory, requiring ECPs to recall information, and therefore was subjected to selection and recall bias.

Keywords: misuse of antibiotics; orthokeratology; contact lens; microbial keratitis; questionnaire

1. Introduction

Orthokeratology (ortho-k) employs specially designed rigid gas-permeable contact lenses, worn overnight to correct myopic refractive error during the day through temporary molding of the cornea [1]. The mechanism underlying the corneal reshaping process mainly involves producing hydraulic force beneath the lens, leading to the redistribution of corneal epithelial cells [2], resulting in central flattening and mid-peripheral steepening [3]. In addition to this vision-correcting function, ortho-k has been shown to slow myopia progression in children and juveniles. Studies have shown its myopia control efficacy, based on the percentage of axial elongation reduction, to range from 36% to 46% [4–7].

In the last decade, ortho-k has become the most accessible and popular myopia control modality in China, with over 1.5 million users in 2016 [8]. The number of eye care practitioners (ECPs) also increased dramatically to meet the huge demand for ortho-k. The International Academy of Orthokeratology Asia (IAOA) trained over 10,000 ECPs and issued over 1700 memberships between 2012 and 2016 in Mainland China [8]. Ophthalmologists and optometrists holding a university degree or diploma working in the same clinical setting with ophthalmologists are considered to be the best qualified professionals to fit ortho-k lenses and to prescribe antibiotic eye drops associated with ortho-k lens wear. Other professionals, such as nurses and opticians, are occasionally involved in ortho-k lens practice, most frequently in private clinics and distributor fitting centers, under the supervision of ophthalmologists. In the current study, these professionals are all listed as ECPs.

The safety of overnight contact lens therapy is an important concern for ECPs, especially when this treatment targets children and juveniles. As overnight ortho-k is associated with reduced tear exchange and oxygen tension at the ocular surface, it is suggested that microorganisms build up during sleep and increase the likelihood of microbial keratitis (MK) [9]. Although the overall safety of overnight ortho-k is reported to be good, with the incidence of MK being similar to that of other overnight or extended wear contact lens modalities [10], its consequences can be sight threatening. Typically, ortho-k-related MK occurs 5 to 15 months after the commencement of treatment [10]. Early diagnosis and treatment of MK are extremely important for prognosis, and correct antibiotic use is essential for the effective management of the infection. The most commonly prescribed antibiotic eye drops, unlike the long-favored triple-agent formulation of polymyxin, neomycin, and bacitracin, contain either 0.3% tobramycin (aminoglycoside) or 0.5% levofloxacin (fluoroquinolone).

A recent survey used a questionnaire to explore the antibiotic (eye drops) prescribing habits of ECPs who fit ortho-k contact lenses in Mainland China. The results indicated that the misuse or improper use of antibiotic eye drops is common among ECPs [11]. Therefore, guidelines based on best practice and taking into account deficiencies and errors identified from the survey, were developed and issued to ortho-k ECPs nationwide via an official online account [11]. The reinforcement of the importance of proper antibiotic use in ortho-k practice lasted for six months, mainly through online education.

As these guidelines were promulgated for six months, the purpose of this study was to investigate their effect on knowledge, attitude, and behavioral change in antibiotic use among ortho-k ECPs in China. Further issues identified from this survey may provide insights into a more efficient means of practitioner education, not only for ECPs, but also for other medical practitioners.

2. Results

Over 4000 subscribers of the IAOA official online account received the invitation to complete the questionnaire. A total of 548 recipients completed and returned the questionnaire. The respondents were composed of ophthalmologists (20.9%), optometrists (65.5%), and others (13.6%). The details of the respondents are shown in Table 1. Respondents' knowledge, attitude, and behaviors regarding antibiotic use are shown in Table 2 and Figures 1–3.

Table 1. Demographics of respondents (n = 548).

	Frequency (%)
1. Occupation	
Ophthalmologist	114 (20.9)
• Corneal specialist	7 (1.3)
• Refractive surgeon	8 (1.5)
• Medical doctor conducting optometry (non-surgical)	99 (18.1)
Optometrist (degree or diploma)	359 (65.5)
Other (nurse, optician, etc.)	75 (13.6)
2. Clinical setting	
General hospital	142 (25.9)
Ophthalmic specialty hospital	120 (21.9)
Private optometry clinic	144 (26.3)
Ortho-k distributor fitting center	142 (25.9)
3. Level of practice	
Provincial level	111 (20.3)
Municipal level	307 (56.0)
County level	55 (10.0)
Others	75 (13.7)
4. Age (years)	
<25	77 (14.1)
25–30	174 (31.8)
31–35	124 (22.6)
36–40	76 (13.9)
>40	97 (17.7)
5. Sex	
Male	143 (26.1)
Female	405 (73.9)

Table 2. Respondents' knowledge and attitudes towards antibiotic eye drop use in orthokeratology practice (n = 548) (%).

Do You Agree with the Statement	Strongly Agree	Agree	Not Sure	Disagree	Strongly Disagree
1. "antibiotic eye drops may be used prophylactically before or after commencement of treatment to prevent corneal infection"?	7 (1.3)	42 (7.7)	57 (10.4)	321 (58.6)	121 (22.0)
2. "when bacterial keratitis is suspected, patients do not have to stop lens wear, but re-enforcement of lens care routines and use of broad-spectrum antibiotics is necessary"?	4 (0.7)	16 (2.9)	24 (4.4)	168 (30.7)	336 (61.3)
3. "avoid dispensing antibiotic eye drops to patients for emergency use (if unavoidable, dispense together with clear written instructions)"?	98 (17.9)	209 (38.1)	41 (7.5)	183 (33.4)	17 (3.1)
4. "it is very important to properly use antibiotic eye drops"?	361 (65.9)	128 (23.4)	26 (4.7)	28 (5.1)	5 (0.9)
5. "the article and the guidelines are useful"?	342 (62.4)	189 (34.5)	12 (2.2)	3 (0.5)	2 (0.4)
6. "I will consider more carefully when using antibiotics after reading the guidelines"?	375 (68.4)	168 (30.7)	4 (0.7)	1 (0.2)	0 (0)

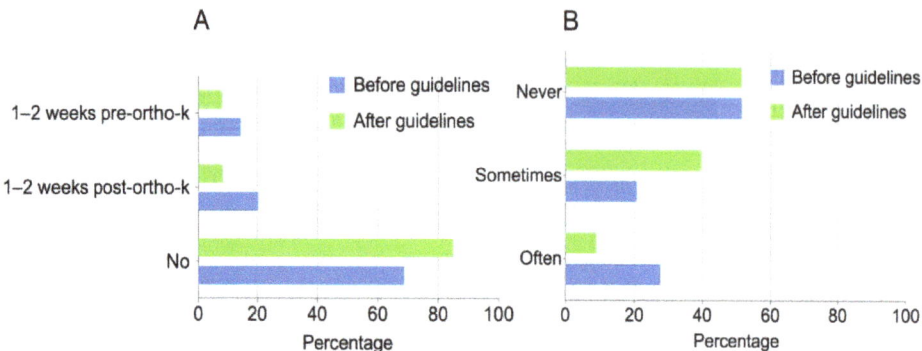

Figure 1. Average frequencies (%), before and after the administration of guidelines, of (**A**) prescription of prophylactic antibiotic eye drops before and after the commencement of ortho-k treatment; (**B**) use of antibiotic eye drops for wetting fluorescein strips during ortho-k lens fitting.

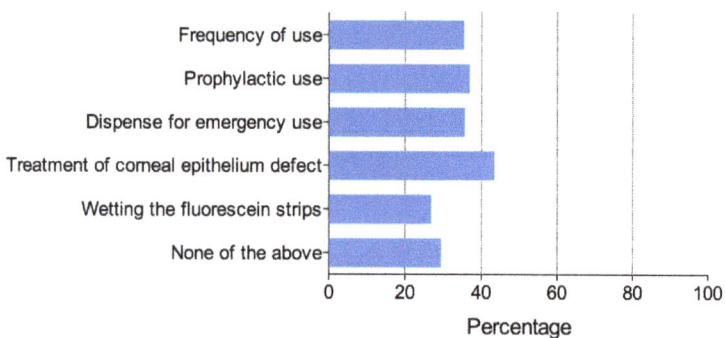

Figure 2. Which aspects of your practice did the article and guidelines change, with respect to the use of antibiotic eye drops in orthokeratology therapy?

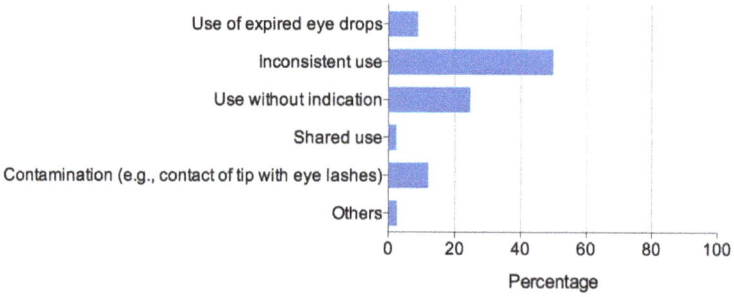

Figure 3. What is the most common misuse of antibiotic eye drops in your orthokeratology patients?

The implementation of the guidelines on the use of antibiotics significantly improved the overall behavior of the ECPs. A total of 39.2% of the respondents of the current study presented the appropriate use of antibiotic eye drops in all three aspects surveyed, compared to 25.9% in the previous study ($p < 0.001$) [11]. The number of antibiotic eye drop prescriptions (based on estimation of purchases) decreased in the preceding six months, with a higher proportion of the respondents reporting less frequent use (i.e., in 0–20% of their patients) than before ($p = 0.027$). A significantly higher proportion of the respondents (84.5%) did not use antibiotics prophylactically in ortho-k patients, compared to the previous survey (67.6%) ($p < 0.001$) (Figure 1A) [11]. A total of 48.5% of the respondents used antibiotic eye drops to wet fluorescein strips, which did not differ from the 48.5% reported in the previous survey. However, when considering the frequency of this misuse, considerably fewer respondents reported "always" or "often" using antibiotic eye drops to wet fluorescein strips (9.0%), as compared to the previous survey (27.7%) ($p < 0.001$) (Figure 1B) [11]. A total of 36.1% of the respondents did not give clear written instructions along with the dispensing of antibiotics to patients, showing no difference compared to the previous survey (40.7%) ($p = 0.132$).

A total of 215 of the respondents (39.2%) used antibiotic eye drops properly in all three aspects surveyed. Their overall knowledge of antibiotic use was significantly better than their counterparts who misused antibiotics in at least one of the behavior-related questions. For example, for the question "antibiotic eye drops may be used prophylactically before or after commencement of ortho-k treatment to prevent corneal infection", 90.6% of the respondents who used antibiotics properly "disagreed" or "strongly disagreed" with this statement, compared to 75.7% of their counterparts ($p < 0.001$). Additionally, for the question "avoid dispensing antibiotic eye drops to patients for emergency use (if unavoidable, dispense together with clear written instructions)", 75.3% of the respondents who used antibiotics properly "agreed" or "strongly agreed" with this statement, compared to 46.4% of their counterparts ($p < 0.001$).

The overall attitude towards the proper use of antibiotics was positive among all the respondents. A total of 89.3% of the respondents "agreed" or "strongly agreed" that it is very important to properly use antibiotic eye drops in ortho-k treatment. The vast majority (96.9%) of the respondents "agreed" or "strongly agreed" that the article and the guidelines were useful. A total of 99.1% of all the respondents "agreed" or "strongly agreed" that they were more careful when considering the use of antibiotics in ortho-k treatment after reading the guidelines. For the question "in what aspects did the article and the guidelines change my practice", the greatest change in behavior was "in the treatment of corneal epithelium defects after ortho-k treatment" (43.4%), and the least change was "in wetting the fluorescein strip using antibiotic eye drops during ortho-k lens fitting" (26.8%) (Figure 2).

The most common misuse of antibiotic eye drops by the patients was "applying eye drops irregularly (period of use or frequency)" (49.8%), followed by "using when not indicated (e.g., allergic conjunctivitis or dry eye)" (24.6%) and "contaminating eye drops via contact of the bottle tip with the eyelashes" (11.9%) (Figure 3). The respondents who used antibiotics properly reported significantly fewer of their patients misusing antibiotics than the ECPs who did not ($p < 0.001$).

Overall, occupation, age, and clinical setting all significantly affected the likelihood of respondents' using antibiotics correctly (Table 3). Significantly higher chances of misusing antibiotics, especially using antibiotic eye drops to wet fluorescein strips, and dispensing antibiotics to patients for emergency use without giving clear written instructions were found among professionals (e.g., nurses and opticians) other than ophthalmologists and optometrists (both $p < 0.05$). Compared to younger respondents (<25 years), older ECPs appeared to be more likely to use antibiotics correctly. Younger respondents (<25 years) were more likely to wet the fluorescein strip with antibiotic eye drops and to dispense antibiotics to patients for emergency use without giving clear written instructions (both $p < 0.05$). Compared to other clinical settings, ECPs in the distributor fitting centers were more likely to use antibiotics improperly, including wetting the fluorescein strip with antibiotic eye drops and dispensing antibiotics to patients for emergency use without giving clear written instructions (both $p < 0.05$). However, they were less likely to use antibiotics prophylactically before or after the commencement of ortho-k treatment ($p < 0.05$).

Table 3. The results of logistic regression models of respondents' demographics on overall proper antibiotic eye drop use.

		OR (95% CI)	p
Occupation	Ophthalmologist	Referent	
	Optometrist (degree or diploma)	0.68 (0.44–1.05)	0.08
	Other (nurse, optician, etc.)	0.23 (0.11–0.48)	<0.001
Clinical setting	Private optometry clinic	Referent	
	Ophthalmic specialty hospital	1.08 (0.66–1.78)	0.75
	General hospital	1.02 (0.64–1.65)	0.92
	Distributor fitting center	0.27 (0.15–0.48)	<0.001
Practice level	Provincial	Referent	
	Municipal	0.77 (0.49–1.22)	0.27
	County	1.01 (0.52–1.99)	0.97
	Others	1.12 (0.61–2.05)	0.72
Age	<25	Referent	
	25–30	2.94 (1.5–5.74)	<0.001
	31–35	2.61 (1.3–5.27)	0.01
	36–40	2.87 (1.35–6.12)	0.01
	>40	2.54 (1.22–5.26)	0.01
Sex	Male	Referent	
	Female	1.46 (0.96–2.22)	0.08

3. Discussion

This study found that, compared to the previous survey, before the development and dissemination of guidelines on the use of antibiotic eye drops, the overall behavior of the respondents had improved significantly since the implementation of the guidelines six months earlier, although the prescribing habits of some ECPs required further improvement.

The greatest change occurred in the prophylactic use of antibiotics before or after ortho-k treatment, as evidenced by a higher proportion (84.5%) of respondents not using antibiotics prophylactically, as compared to the previous survey (67.6%), and also by a decreased number of antibiotic prescriptions in the preceding six months. The most common reasons for using prophylactic antibiotics among ECPs, as reflected by the current survey were: 1. to prevent infection in case of corneal injury during the early phase of ortho-k lens wear; 2. to reduce the discomfort of ortho-k lens wear; 3. to reduce ocular discharge during the early adaptation period of ortho-k; 4. to prevent infection in all sorts of contact lens wear. Since the eyes of healthy young ortho-k patients do not usually carry a heavy burden of microorganisms, there is no evidence that antibiotic eye drops can prevent contact lens-related MK. Notably, ortho-k-induced MK typically occurs five to 15 months after the commencement of lens wear [10], and is most likely to be induced by improper lens care, such as inadequate hand washing, rinsing, storing contact lenses in tap water, or poor hygiene with lens cases and suction holders [12,13]. In addition, symptoms, including foreign body sensation, and clinical signs, such as increased discharge in the early phase of ortho-k lens wear, can be completely normal, so antibiotics for these problems are not only useless, but also add to the risk of bacterial resistance to antibiotics.

Surprisingly, the implementation of the guidelines barely had any effect on the ECPs' habits of prescribing antibiotics to ortho-k patients for use at their own discretion during emergencies. A total of 36.1% of the respondents did not give clear written instructions when dispensing antibiotics to patients. The respondent's occupation (most frequent in "others", such as nurse and optician), the ECP's age (most frequent in those <25 years), and clinical setting (most frequent in distributor fitting centers) were identified as risk factors for this behavior. The most common reasons include: 1. patients live far away or overseas, and it is inconvenient to come to the clinic during an emergency; 2. patients have a heavy academic burden and are not compliant with routine follow-up visits; 3. patients may suffer seasonal allergic conjunctivitis and may need antibiotics "just in case"; 4. patients ask for antibiotics. It is evidenced by our study that the most common misuse of antibiotics among ortho-k patients is to use them irregularly (e.g., period of use or frequency) (49.8%) and to use them when not indicated (e.g., allergic conjunctivitis or dry eye) (24.6%). While it can be argued that medical resources are unevenly distributed across different regions in China and that patients may not have access to their ECP during an emergency, it is doubtful that dispensing antibiotics to patients is justified or ethical. An emergency contact number should be provided to patients along with clear instructions as to when to call [14]. For patients with poor compliance with follow-up visits, there is no reason to believe that they can use antibiotics correctly when indicated and, considering the potential risk arising from non-compliance with treatment, they are probably not suitable candidates for ortho-k treatment. Currently, there are several novel non-contact lens myopia control modalities other than ortho-k, including specially designed spectacles [15,16] and low-concentration atropine [17,18], which could be considered when compliance is an issue for patients.

One of the disturbing behaviors that still needs urgent attention is the use of antibiotic eye drops to wet fluorescein strips during ortho-k lens fitting. It is worrisome that no change has been observed in the proportion of ECPs conducting this practice since the administration of the guidelines, although significantly fewer respondents reported that they "always" or "often" use antibiotic eye drops to wet fluorescein strips (9.0%), as compared to the previous survey (27.7%). In concordance with this finding is that the fewest respondents reported a change in this behavior when responding to the question "In what aspects did the article and the guidelines change your practice". The respondent's occupation (most frequent in "others", such as nurse and optician), the ECP's age (most frequent in <25 years) and clinical setting (most frequent in distributor fitting centers) were the risk factors identified for this behavior. The most common reasons for this behavior are that: 1. antibiotic eye drops are easily accessible in the clinic; 2. no other solutions, such as saline solutions, are available in the clinic; 3. antibiotic eye drops may decrease the irritation and prevent infection during ortho-k lens fitting. These results indicated that the clinical setting plays a major role in the regulation of antibiotic use in ortho-k lens fitting. Hospitals and clinics should restrict the use of antibiotics, provide ortho-k ECPs with antibiotic stewardship programs (especially for younger ECPs and ECPs identified as other than ophthalmologists and optometrists working in distributor fitting centers), and, most importantly, provide non-preserved saline solution instead of antibiotic eye drops to wet fluorescein strips during ortho-k lens fitting. Trial lens disinfection should also be emphasized in clinical settings to address ECPs' concerns about cross-infection among ortho-k patients.

Generally, the overall attitude towards proper antibiotic use was positive. Therefore, the misuse of antibiotics by some respondents cannot be solely attributable to their indifference to the guidelines. The guidelines have been distributed on various online platforms and read by ECPs over 5000 times. Since the correlation between knowledge and behavior was found to be strong in this study, we are tempted to conclude that the efficiency of education mainly through online social media was insufficient. Reinforcement of guidelines in off-line conferences and continuing education programs are also suggested. Interestingly, it was noted that the respondents who used antibiotics properly reported fewer of their patients misusing antibiotics than the ECPs who did not. This finding indicated that ECPs' knowledge about antibiotics might impact their behavior, which in turn influences their patients' compliance with antibiotic use.

One important issue identified in the current study that needs urgent attention is that the overall antibiotic misuse is most common among professionals other than ophthalmologists and optometrists, i.e., nurses and opticians. They are less likely to receive full-term medical training or optometric education, not to mention fitting specialty contact lenses such as ortho-k. Therefore, not only should the guidelines on antibiotic use be delivered to them in a timely manner, but regulations should also be reinforced to limit their prescription of this specialized therapy.

One of the limitations of this study was that the two surveys were not conducted on the same cohort, since the first one was performed anonymously [11]. Therefore, the data collected could not be used to directly compare the behaviors before and after the administration of guidelines in individuals, but rather, in a group of practitioners. However, the respondents were evenly distributed across all clinical settings, indicating that the sample was representative in terms of medical practice. As with all self-administered questionnaires, there is always a question of the truthfulness of the responses. A survey consisting of all mandatory questions may contribute to a selection bias in the participants of the survey. Additionally, the surveys were based on impressions or perceptions of how the prescriber uses antibiotics and did not always correspond to prescribing or dispensing data with respect to the best practices in antimicrobial use. However, even if a respondent did not answer all questions accurately, just responding to the questions would result in a reminder of the correct procedures. As such, surveys can act as a tool for the reinforcement of good practice.

4. Materials and Methods

4.1. Preparation of the Questionnaire

A questionnaire (see Supplementary Material S1), comprising 21 questions (16 related to the knowledge, attitude, and behavior in antibiotic prescription and five related to demographics), was prepared and reviewed by five contact lens specialists from four top-ranking national ophthalmology departments. The questionnaires were prepared initially in English, translated into Chinese, and then translated back into English to check for accuracy. The respondents were requested to complete the questions in the Chinese version. All the questions in this survey were mandatory. The questionnaire was then sent to 35 ECPs involved in ortho-k lens treatment nationwide (not included in the final analysis) for validation and to calculate the internal consistency of the questionnaire. The Cronbach's alpha coefficient was 0.75, which indicated an acceptable level of the questionnaire's reliability. The three key questions regarding habitual use of antibiotic eye drops were:

- Do you prescribe prophylactic antibiotic eye drops before or after fitting ortho-k?
- Do you dispense antibiotic eye drops to patients for emergency use (at patients' discretion)?
- Do you use antibiotic eye drops to wet fluorescein strips during ortho-k lens fitting?

4.2. Definition of Proper Antibiotic Use

To further determine which group of ECPs was more likely to misuse or overuse antibiotics, and whether the circulated guidelines had had any effect on the habitual use of antibiotics by ECPs, options for the above-mentioned questions were categorized based on the definitions of proper use of antibiotics below:

- Do not use prophylactic antibiotic eye drops before or after fitting ortho-k lenses;
- Do not dispense antibiotic eye drops to patients for emergency use (when essential, dispense to patients along with written instructions);
- Do not use antibiotic eye drops to wet fluorescein strips during ortho-k lens fitting.

4.3. Distribution of Questionnaires

The questionnaires were distributed through the official online account of the International Academy of Orthokeratology Asia (IAOA), which has over 4000 members listed as ortho-k practitioners throughout China. The survey was conducted over June and July 2020. The inclusion criteria comprised ECPs, including ophthalmologists, optometrists, and other professionals, such as nurses and opticians, who had been involved in ortho-k lens fitting and patient care.

4.4. Statistical Analyses

Data concerning prescribing habits (16 questions) and demographics of the ECPs (five questions) were expressed as percentages. Data of the three key prescribing habits were compared to those of the previous study using χ^2 tests. Using the definition of proper antibiotic use, ECPs were further divided into groups, and the differences between the groups were tested using χ^2 tests for categorical variables. Stepwise logistic regression models were used to assess the relationship between prescribing habits and demographics. Two-sided values of $p < 0.05$ were deemed statistically significant. All statistical analyses were performed using software R (version 4.0.5).

5. Conclusions

The current study revealed that the overall behavior of the ortho-k ECPs significantly improved after the nationwide implementation of guidelines on antibiotic use, although some ECPs' prescribing habits require further improvement. Since the ECPs' behavior was significantly affected by their knowledge, the effective delivery of education is the key to success: it should not only be delivered online, but also off-line to enhance its efficacy. Clinical settings, such as hospitals and clinics, can also play an important role by providing ECPs with non-preserved saline solution to replace antibiotic eye drops in ortho-k lens fitting, and by introducing an antibiotic stewardship program, especially for younger ECPs. Rigorous clinical governance measures, particularly at the distributor fitting centers, are strongly recommended with regard to ortho-k practice and antibiotic use.

Supplementary Materials: The following supporting information can be downloaded at: https://www.mdpi.com/article/10.3390/antibiotics11020179/s1, File S1: Questionnaire.

Author Contributions: Conceptualization, Z.C., J.W. and P.C.; design of questionnaire, statistical analyses, J.W.; investigation, Z.C. and J.W.; writing—original draft preparation, Z.C.; writing—review and editing, J.W., J.J., B.Y. and P.C.; supervision, P.C. All authors have read and agreed to the published version of the manuscript.

Funding: This research received no external funding.

Informed Consent Statement: Consent forms are not needed, as this is a survey recalling information. No human subjects were involved.

Data Availability Statement: Data are available upon request.

Acknowledgments: The authors thank Maureen Boost for her help with the editing of the manuscript.

Conflicts of Interest: The authors declare no conflict of interest.

Ethics Statement: This study was conducted in accordance with the Declaration of Helsinki, and approved by the Ethics Committee of Fudan University Eye and ENT Hospital (protocol code 2021086, 8 June 2021).

References

1. Nichols, J.J.; Marsich, M.M.; Nguyen, M.; Barr, J.T.; Bullimore, M.A. Overnight orthokeratology. *Optom. Vis. Sci.* **2000**, *77*, 252–259. [CrossRef]
2. Alharbi, A.; Swarbrick, H. The effects of overnight orthokeratology lens wear on corneal thickness. *Investig. Ophthalmol. Vis. Sci.* **2003**, *44*, 2518–2523. [CrossRef]
3. Maseedupally, V.; Gifford, P.; Lum, E.; Swarbrick, H. Central and paracentral corneal curvature changes during orthokeratology. *Optom. Vis. Sci.* **2013**, *90*, 1249–1258. [CrossRef] [PubMed]
4. Cho, P.; Cheung, S.W. Retardation of myopia in orthokeratology (ROMIO) study: A 2-year randomized clinical trial. *Investig. Ophthalmol. Vis. Sci.* **2012**, *53*, 7077–7085. [CrossRef]
5. Cho, P.; Cheung, S.W.; Edwards, M. The longitudinal orthokeratology research in children (LORIC) in Hong Kong: A pilot study on refractive changes and myopic control. *Curr. Eye Res.* **2005**, *30*, 71–80. [CrossRef] [PubMed]
6. Hiraoka, T.; Kakita, T.; Okamoto, F.; Takahashi, H.; Oshika, T. Long-term effect of overnight orthokeratology on axial length elongation in childhood myopia: A 5-year follow-up study. *Investig. Ophthalmol. Vis. Sci.* **2012**, *53*, 3913–3919. [CrossRef] [PubMed]
7. Walline, J.J.; Jones, L.A.; Sinnott, L.T. Corneal reshaping and myopia progression. *Br. J. Ophthalmol.* **2009**, *93*, 1181–1185. [CrossRef] [PubMed]
8. Xie, P.; Guo, X. Chinese experiences on orthokeratology. *Eye Contact Lens* **2016**, *42*, 43–47. [CrossRef] [PubMed]
9. Hoddenbach, J.G.; Boekhoorn, S.S.; Wubbels, R.; Vreugdenhil, W.; Van Rooij, J.; Geerards, A.J. Clinical presentation and morbidity of contact lens-associated microbial keratitis: A retrospective study. *Graefes Arch. Clin. Exp. Ophthalmol.* **2014**, *252*, 299–306. [CrossRef] [PubMed]

10. Bullimore, M.A.; Sinnott, L.T.; Jones-Jordan, L.A. The risk of microbial keratitis with overnight corneal reshaping lenses. *Optom. Vis. Sci.* **2013**, *90*, 937–944. [CrossRef] [PubMed]
11. Chen, Z.; Jiang, J.; Xu, J.; Yang, X.; Yang, Y.; Wang, K.; Song, H.; Yang, B.; Cho, P. Antibiotic eye drops prescription patterns by orthokeratology practitioners in china and the development of antibiotic usage guidelines. *Contact Lens Anterior Eye* **2021**, *44*, 101354. [CrossRef] [PubMed]
12. Watt, K.G.; Swarbrick, H.A. Trends in microbial keratitis associated with orthokeratology. *Eye Contact Lens* **2007**, *33*, 373–377; discussion 382. [CrossRef] [PubMed]
13. Cho, P.; Boost, M.; Cheng, R. Non-compliance and microbial contamination in orthokeratology. *Optom. Vis. Sci.* **2009**, *86*, 1227–1234. [CrossRef] [PubMed]
14. Cho, P.; Cheung, S.W.; Mountford, J.; White, P. Good clinical practice in orthokeratology. *Contact Lens and Anterior Eye* **2008**, *31*, 17–28. [CrossRef] [PubMed]
15. Zhang, H.Y.; Lam, C.S.; Tang, W.C.; Leung, M.; To, C.H. Defocus incorporated multiple segments spectacle lenses changed the relative peripheral refraction-a 2-year randomized clinical trial (DIMS). *Investig. Ophthalmol. Vis. Sci.* **2020**, *61*, 53. [CrossRef] [PubMed]
16. Bao, J.; Yang, A.; Huang, Y.; Li, X.; Pan, Y.; Ding, C.; Lim, E.W.; Zheng, J.; Spiegel, D.P.; Drobe, B.; et al. One-year myopia control efficacy of spectacle lenses with aspherical lenslets. *Br. J. Ophthalmol.* 2021, *online ahead of print*. [CrossRef] [PubMed]
17. Chia, A.; Chua, W.H.; Wen, L.; Fong, A.; Goon, Y.Y.; Tan, D. Atropine for the treatment of childhood myopia: Changes after stopping atropine 0.01%, 0.1% and 0.5%. *Am. J. Ophthalmol.* **2014**, *157*, 451–457.e1. [CrossRef] [PubMed]
18. Yam, J.C.; Li, F.F.; Zhang, X.; Tang, S.M.; Yip, B.H.K.; Kam, K.W.; Ko, S.T.; Young, A.L.; Tham, C.C.; Chen, L.J.; et al. Two-year clinical trial of the low-concentration atropine for myopia progression (LAMP) study: Phase 2 report. *Ophthalmology* **2019**, *127*, 910–919. [CrossRef] [PubMed]

Article

Antibiotics Used in Empiric Treatment of Ocular Infections Trigger the Bacterial Rcs Stress Response System Independent of Antibiotic Susceptibility

Nathaniel S. Harshaw, Nicholas A. Stella, Kara M. Lehner †, Eric G. Romanowski, Regis P. Kowalski and Robert M. Q. Shanks *

Charles T. Campbell Ophthalmic Microbiology Laboratory, Department of Ophthalmology, University of Pittsburgh School of Medicine, Pittsburgh, PA 15213, USA; nah126@pitt.edu (N.S.H.); nas92@pitt.edu (N.A.S.); kmlehner@udel.edu (K.M.L.); romanowskieg@upmc.edu (E.G.R.); kowalskirp@upmc.edu (R.P.K.)
* Correspondence: shanksrm@upmc.edu; Tel.: +1-412-647-3537
† Current address: Department of Biological Sciences, University of Delaware, Newark, DE 19803, USA.

Abstract: The Rcs phosphorelay is a bacterial stress response system that responds to envelope stresses and in turn controls several virulence-associated pathways, including capsule, flagella, and toxin biosynthesis, of numerous bacterial species. The Rcs system also affects antibiotic tolerance, biofilm formation, and horizontal gene transfer. The Rcs system of the ocular bacterial pathogen *Serratia marcescens* was recently demonstrated to influence ocular pathogenesis in a rabbit model of keratitis, with Rcs-defective mutants causing greater pathology and Rcs-activated strains demonstrating reduced inflammation. The Rcs system is activated by a variety of insults, including β-lactam antibiotics and polymyxin B. In this study, we developed three luminescence-based transcriptional reporters for Rcs system activity and used them to test whether antibiotics used for empiric treatment of ocular infections influence Rcs system activity in a keratitis isolate of *S. marcescens*. These included antibiotics to which the bacteria were susceptible and resistant. Results indicate that cefazolin, ceftazidime, polymyxin B, and vancomycin activate the Rcs system to varying degrees in an RcsB-dependent manner, whereas ciprofloxacin and tobramycin activated the promoter fusions, but in an Rcs-independent manner. Although minimum inhibitory concentration (MIC) analysis demonstrated resistance of the test bacteria to polymyxin B and vancomycin, the Rcs system was activated by sub-inhibitory concentrations of these antibiotics. Together, these data indicate that a bacterial stress system that influences numerous pathogenic phenotypes and drug-tolerance is influenced by different classes of antibiotics despite the susceptibility status of the bacterium.

Keywords: Enterobacterales; keratitis; infection; cornea; bacteria; stress response system; antibiotic

1. Introduction

Serratia marcescens is a Gram-negative pathogen from the order Enterobacterales that causes contact lens-associated keratitis in healthy patients [1–3] and a wide variety of nosocomial infections in the immune compromised, such as ventilator-associated pneumonia and sepsis in adults and neonates [4,5]. *S. marcescens* isolates are typically resistant to antibiotics of the macrolide, tetracycline, β-lactam, and narrow spectrum cephalosporin classes due to expression of efflux pumps and β-lactamases [6]. However, they are generally susceptible to aminoglycoside, third generation cephalosporin, and fluoroquinolone antibiotics [6,7].

The Rcs stress response system has been found in bacteria from the Enterobacterales including, but not limited to, numerous pathogens, such as *Escherichia coli*, *Klebsiella* species, *Proteus mirabilis*, *Salmonella enterica*, and *Yersinia pestis* [8]. The core Rcs system (Figure 1) is a complex signal transduction cascade composed of a variety of components that include outer membrane protein RcsF, inner membrane protein IgaA, two histidine kinase-related proteins, RcsC and RcsD, and the response regulator transcription factor RcsB [8]. Rcs signaling occurs in response to cell envelope stresses, such as defects in peptidoglycan and lipopolysaccharide (LPS) structure, perturbations of the outer membrane β-barrel protein assembly complex, and lipoprotein trafficking [8,9]. Antimicrobials known to activate the Rcs system, mostly from studies with *E. coli* and *S. enterica*, include polymyxin B [10] and other antimicrobial peptides [11], and cell wall-targeting β-lactam and cephalosporin antibiotics [9,12]. However, this has not been tested in ocular pathogens such as *S. marcescens*.

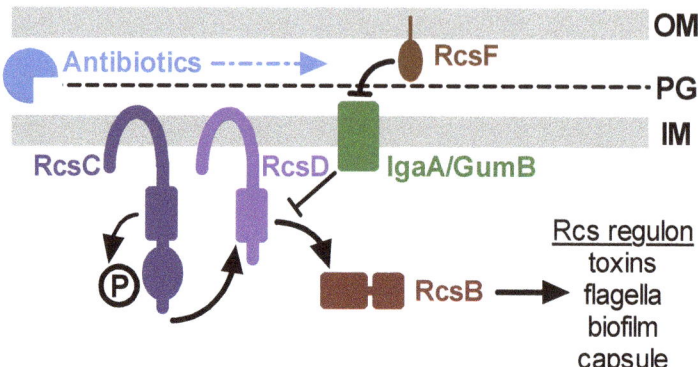

Figure 1. Model for antibiotic activation of the Rcs system. This simplified depiction of the core Rcs system shows the major components required for Rcs function. The Rcs system is a complex phosphorelay signal transduction system that regulates the transcription of many genes through control of the RcsB transcription factor. The IgaA/GumB inner membrane protein blocks Rcs activity under non-stressful conditions. Envelope stress by antibiotics, transmitted by RcsF, prevents IgaA/GumB inhibition of RcsC-D and allows RcsB-mediated transcription. Mutation of *igaA/gumB* constitutively derepresses the Rcs transcriptional cascade, and mutation of *rcsB* prevents Rcs system function. This model predicts that Rcs activation by antibiotics can stimulate pathogenesis and antibiotic tolerance phenotypes. OM: outer membrane; PG: peptidoglycan; IM: inner membrane.

Importantly, the Rcs system has been shown to contribute to antibiotic tolerance by a number of bacteria, with Rcs system-defective mutants being more susceptible to penicillin and cephalosporin antibiotics for *E. coli* [12] and polymyxin B for *E. coli* [13] and *S. enterica* [10]. Similarly, induced expression of *rcsB* or expression of alleles that increase Rcs activity conferred increased tolerance to β-lactam and cephalosporin antibiotics for *E. coli* [12,14]. A major mechanism used by bacteria to increase antibiotic tolerance is biofilm formation [15]. The Rcs system plays a positive role in *S. marcescens* biofilm formation under high sheer conditions by promoting capsular polysaccharide synthesis [16]. A similar role for the Rcs system in *E. coli* and *S. enterica* biofilm formation has been described [17]. Beyond antibiotic tolerance, a recent study by Smith et al. suggests that Rcs plays a role in the acquisition of genetic elements by *Serratia* sp. 39006 that may contribute to horizontal gene transfer and antibiotic resistance [18].

The *S. marcescens* Rcs system has been shown to regulate synthesis of the ShlA cytolysin [19,20], where it is also a key regulator of capsular polysaccharide and flagella synthesis, as well as the production of a hemolytic biosurfactant [16,21]. Importantly, the Rcs system was shown to be a major regulator of *S. marcescens* ocular pathogenesis [22].

The goal of this study was to evaluate whether antibiotics commonly used topically for empiric treatment of ocular infections activate the bacterial Rcs pathway. In this study, we used antibiotics recommended for the empiric treatment of bacterial keratitis by the American Academy of Ophthalmology [23]. Given the role of the Rcs system in promoting antibiotic tolerance and the regulation of virulence factors, it is possible that activation of this system could influence clinical outcomes for patients infected by the Enterobacterales. To that end, we developed luminescent reporter plasmids for Rcs activity and used them in a keratitis isolate of *S. marcescens* with antibiotics from several classes that are recommended for the treatment of ocular infections.

2. Results

2.1. Generation of Luminescent Reporter Plasmids for Rcs System Activity

In order to conveniently measure Rcs activation, luminescent reporter plasmids were made using Rcs-responsive promoters. GumB, an IgaA ortholog, is a negative regulator of Rcs activity, such that a *gumB* deletion mutant has a highly activated Rcs system [16,19]. Transcriptomic analysis of a Δ*gumB* mutant was used to identify genes that were more highly expressed than in the wild type (to be described elsewhere). Three promoters were cloned upstream of the *luxCDABE* operon on a broad-host range low-copy vector (Figure 2A and Figure S1). The promoters were for the SMDB11_1637, SMDB11_2817, and SMDB11_1194 open reading frames. All of these previously uncharacterized open reading frames bear high similarity to Rcs-regulated genes in other bacteria. SMDB11_1637 is similar to osmotically inducible lipoprotein B (*osmB*), which is positively regulated by the Rcs system in *Erwinia amylovora* [24], *E. coli* [25,26], *P. mirabilis* [27], *S. enterica* [28], and *Yersinia pseudotuberculosis* [29]. SMDB11_1194 is highly similar to *umoD*, which is Rcs-regulated in *P. mirabilis* [27], as is its ortholog YPO1624 in *Y. pseudotuberculosis*. SMDB11_2817 has similarity to *yaaX* from *E. coli* with the DUF2502 domain of unknown function and was identified as an RcsB-regulated gene in *E. coli* [25].

In addition, the *nptII* promoter from the Tn5 transposon was used as a constitutive control promoter to test for theoretical physiological conditions that could interfere with luminescence.

To validate the Rcs system activation of these promoters, they were moved into a contact lens-associated keratitis wild-type (WT) isolate of *S. marcescens*, strain K904, and isogenic mutants with manipulated Rcs systems that confer high (Δ*gumB*) or no (Δ*rcsB* and Δ*gumB* Δ*rcsB*) Rcs activity. Strains and plasmids are listed in Table S1 (Supplementary Materials). Bacteria were grown overnight, and the luminescence was determined as a function of culture optical density (Figure 2). The test strains were previously shown to achieve similar growth levels over the tested time frame [30].

The results (Figure 2) suggested that the promoter activity for each of the genes, except the control *nptII* promoter, was highly increased (>6000-fold) in the Rcs-activated mutant background (Δ*gumB*). Furthermore, a clear reduction in luminescence was observed in the Δ*gumB* Δ*rcsB* double mutant, confirming that the increase observed in the Δ*gumB* mutant was Rcs dependent. There was some expression in the absence of Rcs activity (see Δ*rcsB* mutant), indicating that there is some Rcs-independent expression from these promoters (i.e., other transcription factors may regulate some of these promoters; see discussion). Importantly, the *nptII* promoter showed only a minor but significant change (~2-fold) among the various mutant backgrounds, suggesting that the Rcs system status has little to no impact on the ability of the luminescent reporter system to function. Together, these results indicate that we have identified and cloned three Rcs system-responsive promoters and created reporter constructs to analyze compounds that may influence Rcs-activity.

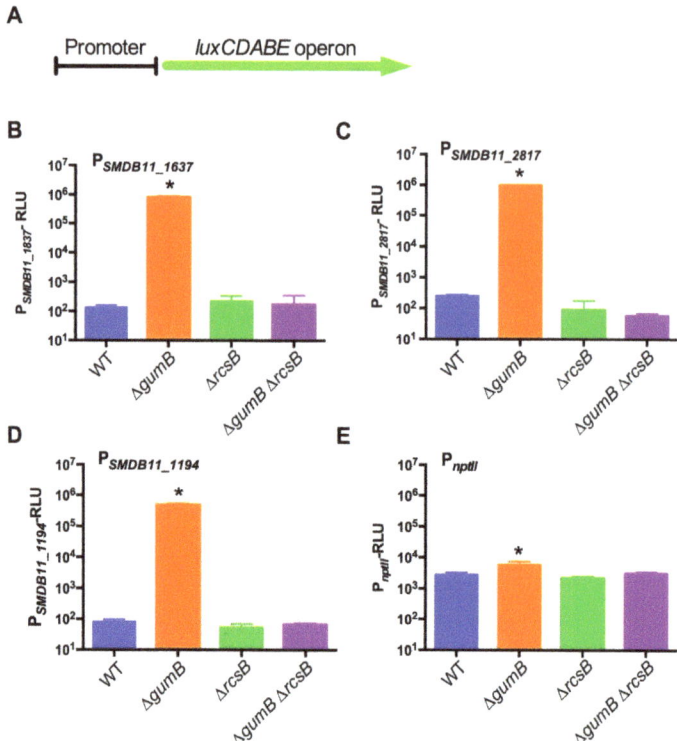

Figure 2. Validation of Rcs-responsive transcriptional reporter plasmids. (**A**). Schematic diagram of a promoter transcriptional fusion to the luminescence-producing *luxCDABE* operon that was cloned into a broad-host range medium-copy plasmid. Four different promoters were evaluated by moving them into *S. marcescens* with normal (WT), hyper-activated (ΔgumB), or defective (ΔrcsB, ΔgumB ΔrcsB) Rcs-systems. (**B–E**). Transcription from the four promoters was measured using a luminometer after the bacteria were grown for 20 h in LB medium (n = 4–6). The luminescence values were normalized by optical density, which was similar for each genotype. The *nptII* promoter is an *E. coli* promoter that was used as a constitutive control. The PSMDB11_1637, PSMDB11_2817, and PSMDB11_1194 promoters were Rcs responsive. The asterisks (*) indicate that the ΔgumB group is statistically different than the other groups, $p < 0.01$. WT: wild type.

2.2. Antibiotics Targeting the Cell Envelope Activate the S. marcescens Rcs System Regardless of the Antibiotic Susceptibility Status of the Bacterium

The antibiotic susceptibility of four antibiotics used in treatment of ocular infections was analyzed: polymyxin B, cefazolin, ceftazidime, and vancomycin (Tables 1 and 2). These target either the peptidoglycan cell wall or the bacterial membrane. Minimum inhibitory concentrations (MICs) for the antibiotics to inhibit *S. marcescens* strain K904 were determined (Table 2). The isolate was susceptible to ceftazidime, but was able to grow at the highest tested concentrations of polymyxin B, cefazolin, and vancomycin. This was a typical pattern for keratitis isolates of *S. marcescens* [7]. Nevertheless, prior to identification of the infecting microbe, any of the antibiotics other than polymyxin B are candidates for empiric therapy for keratitis.

Table 1. Characteristics of antibiotics used in this study.

Antibiotic	Typical Topical Drug Concentration [23]	Corneal Tissue Concentration	Typical Systemic Dose	Peak Serum Concentration (µg/mL)	Antibiotic Concentration Used in This Study (µg/mL)
Cefazolin	50 mg/mL Fortified [23]	NA	1 g (IV) q8h [31]	200 µg/mL [31]	39–1250
Ceftazidime	50 mg/mL Fortified [23]	NA	2 g (IV) q8h [31]	120 µg/mL [31]	39–1250
Ciprofloxacin	3 mg/mL Commercial [23]	9.92 ± 10.99 µg/g [32]	400 mg (IV) q12h [31] 500–750 mg (PO) q12h [31]	4.6 µg/mL [31] (IV) 2.8 µg/mL [31] (PO)	2.5–75
Polymyxin B	0.75–1 mg/mL (7500–10,000 units/mL) Commercial [33]	NA	1.25 mg/kg (IV) q12h (1 mg = 10,000 units) [31]	8 µg/mL [31]	30–10,000
Tobramycin	3 mg/mL Commercial [34] 9–14 mg/mL Fortified [23]	NA	5 mg/kg (IV) q24h [31] or 240 mg (IV) q24h [31] (preferred over q8h dosing)	16–24 µg/mL q24h dosing [31]	8–250
Vancomycin	10–50 mg/mL [23,35]	46.7 µg/g [36]	1 g (IV) q12h [31]	40 µg/mL [31]	39–1250

NA: information not available; IV: intravenous; PO: per os; q8h: every 8 h; q12h: every 12 h, q24h: every 24 h.

Table 2. Antibiotic susceptibility analysis of *S. marcescens* strain K904.

Antibiotic	Class	Target	MIC [a]—WT (µg/mL)	MIC—Δ*rcsB* (µg/mL)	Susceptibility [b]	Rcs-Specific Induction [c]
Cefazolin	Cephalosporin	Cell wall	>256, >256	>256, >256	No	Yes
Ceftazidime	Cephalosporin	Cell wall	0.25, 0.19	0.19, 0.19	Yes	Yes
Ciprofloxacin	Fluoroquinolone	DNA gyrase and topoisomerase IV	0.064, 0.064	0.094, 0.470	Yes	No
Polymyxin B	Polymyxin	Cell membrane	>1024, >1024	>1024, >1024	No	Yes
Tobramycin	Aminoglycoside	Ribosome	2, 2	1.5, 1.5	Yes	No
Vancomycin	Glycopeptide	Cell wall	>256, >256	>256, >256	No	Yes

[a] Minimum inhibitory concentrations (MICs) were determined by E-test; values for two independent tests are shown. [b] Susceptibility status was based on Clinical and Laboratory Standards Institute breakpoints [37]. [c] At least one promoter was activated in the wild type, but none in the Δ*rcsB* mutant.

Polymyxin B was previously shown to activate the Rcs system of *E. coli* and *S. enterica* [11,13]. Unlike these bacteria, the vast majority of *S. marcescens* isolates are resistant to polymyxin B due to a 4-aminoarabinose modification of the lipid A portion of the lipopolysaccharide molecules that populate the outer leaflet of the outer membrane [38]. The K904 strain was evaluated for polymyxin B susceptibility and found to be resistant (MIC > 1024, Table 2). The induction of the Rcs system by polymyxin B in a resistant bacterial species has not been evaluated.

Polymyxin B did not activate the *nptII* promoter in the WT bacteria, as expected (Figure 3A); however, the Rcs-dependent promoters were activated in an antibiotic dose-dependent manner, up to 5–10 fold above the absence of antibiotic (Figure 3B–D). To ensure that the effect was Rcs-dependent, the reporters were tested in an isogenic Δ*rcsB* mutant strain. While there was a less than 2-fold increase in luminescence correlating with the presence of antibiotics, it was not dose dependent in the Δ*rcsB* mutant (Figure 3B–D). These suggest that polymyxin B activates the Rcs system even in a resistant bacterium.

The identical approach was used for three different classes of cell wall-targeting antibiotics. A β-lactam antibiotic, cefazolin, is used to treat Gram-positive bacteria (Table 1). *S. marcescens* strain K904 was resistant to cefazolin (Table 2). Experiments indicated very little induction except in the SMDB11_1194 promoter (Figure 4). Similarly, *S. marcescens* K904 was resistant to the glycopeptide vancomycin (Tables 1 and 2) and was activated by the three Rcs-dependent promoters in the WT but not the Δ*rcsB* mutant (Figure 5). By contrast, *S. marcescens* was susceptible to the cephalosporin ceftazidime (Tables 1 and 2).

Two of the Rcs-dependent promoters were activated by ceftazidime in the WT but not the ΔrcsB mutants (Figure 6).

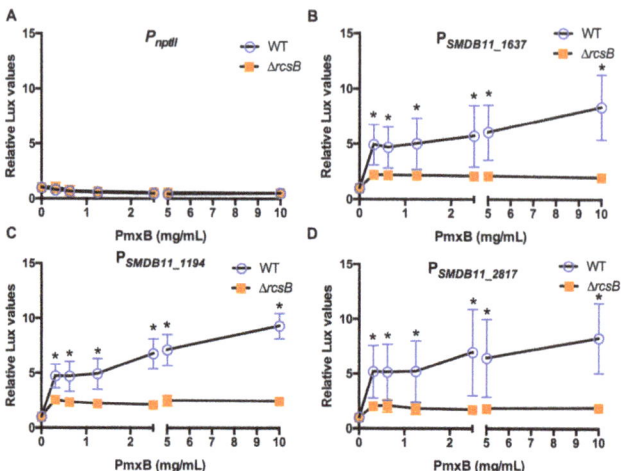

Figure 3. Effect of cell envelope-targeting antibiotic polymyxin B on Rcs-activated promoters (**A–D**). Relative luminescence values were determined by dividing luminescence by optical density after 4 h of antibiotic challenge. The *nptII* promoter (**A**) was unaffected by polymyxin B; however, the Rcs-dependent promoters (**B–D**) were activated to a greater extent in the WT than the Rcs-defective ΔrcsB mutant. Mean and standard deviation are shown (n = 6–9 are shown). Asterisks (*) indicate statistical differences between groups at the indicated concentrations, $p < 0.05$.

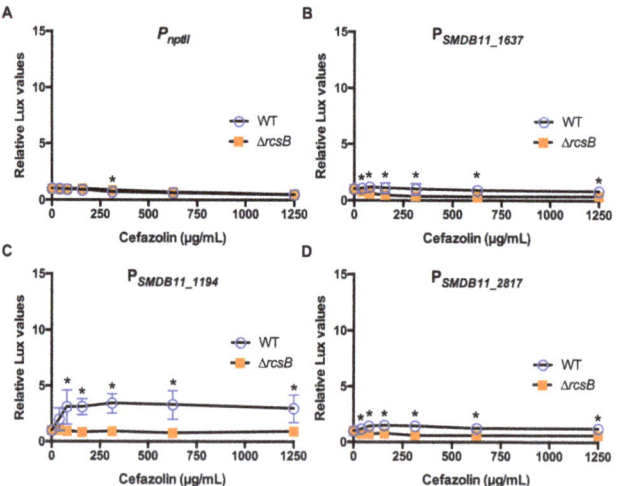

Figure 4. Effect of cell wall activating cefazolin on Rcs-activated promoters (**A–D**). Relative luminescence values were determined by dividing luminescence by optical density after 4 h of antibiotic challenge. The *nptII* promoter (**A**) was unaffected by cefazolin. Only the Rcs-dependent SMDB11_1194 promoter (**C**) was activated to a greater extent in the WT than the Rcs-defective ΔrcsB mutant. Mean and standard deviation are shown (n = 6–9 are shown). Asterisks (*) indicate statistical differences between groups at the indicated concentrations, $p < 0.05$.

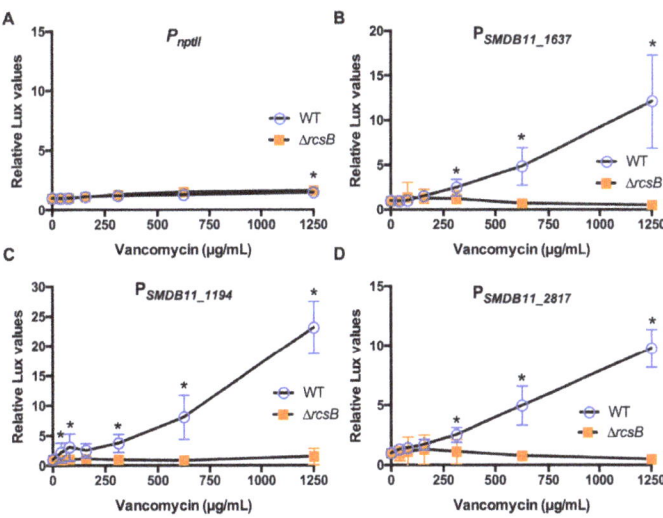

Figure 5. Effect of the cell wall activating antibiotic vancomycin on Rcs-activated promoters (**A–D**). Relative luminescence values were determined by dividing luminescence by optical density after 4 h of antibiotic challenge. The *nptII* promoter (**A**) was unaffected by vancomycin. The experimental promoters (**B–D**) were activated to a greater extent in the WT than the Rcs-defective Δ*rcsB* mutant. Mean and standard deviation are shown (n = 6–9 are shown). Asterisks (*) indicate statistical differences between groups at the indicated concentrations, $p < 0.05$.

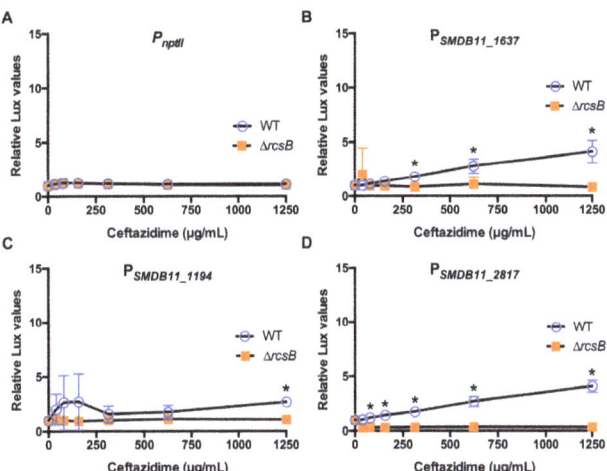

Figure 6. Effect of the cell wall activating antibiotic ceftazidime on Rcs-activated promoters (**A–D**). Relative luminescence values were determined by dividing luminescence by optical density after 4 h of antibiotic challenge. The *nptII* promoter (**A**) was unaffected by ceftazidime. The Rcs-dependent promoters (**B–D**) were activated to a greater extent in the WT than the Rcs-defective Δ*rcsB* mutant. Mean and standard deviation are shown (n = 6–9 are shown). Asterisks (*) indicate statistical differences between groups at the indicated concentrations, $p < 0.05$.

2.3. Non-Cell Envelope-Targeting Antibiotics Activated the Test Promoters in an Rcs-Independent Manner

The same approach used for envelope-targeting antibiotics was used for two non-envelope-targeting antibiotics. Ciprofloxacin is a fluoroquinolone that targets DNA metabolism and is highly effective against Gram-negative ocular pathogens such as *Pseudomonas aeruginosa* and *S. marcescens* (Tables 1 and 2). Figure 7 demonstrates that the three test promoters were highly activated by low concentrations of ciprofloxacin in the WT. However, similar, or even higher levels of expression, were observed in the Rcs-defective mutant, indicating that the activation of the test promoters was Rcs-independent and suggesting that ciprofloxacin does not activate the Rcs system.

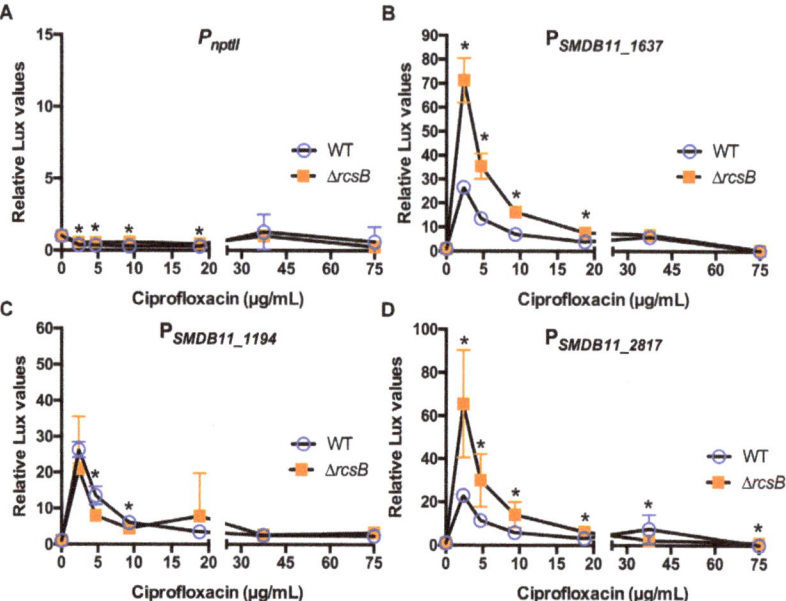

Figure 7. Effect of DNA metabolism-targeting ciprofloxacin on Rcs-activated promoters (**A–D**). Relative luminescence values were determined by dividing luminescence by optical density after 4 h of antibiotic challenge. The *nptII* promoter (**A**) was largely unaffected by ciprofloxacin. The experimental promoters (**B–D**) were activated to an equal or greater extent in the Δ*rcsB* mutant than the WT. Mean and standard deviation are shown (n = 6–9 are shown). Asterisks (*) indicate statistical differences between groups at the indicated concentrations, $p < 0.05$.

The ribosome-targeting aminoglycoside antibiotic tobramycin is used to treat ocular bacterial pathogens (Tables 1 and 2). Data in Figure 8 indicate very little induction of promoter activity by tobramycin except by low induction of the SMDB11_1194 promoter. Slightly higher expression of the promoters was observed in the Δ*rcsB* mutant, suggesting that the promoter transcriptional activation was Rcs-independent.

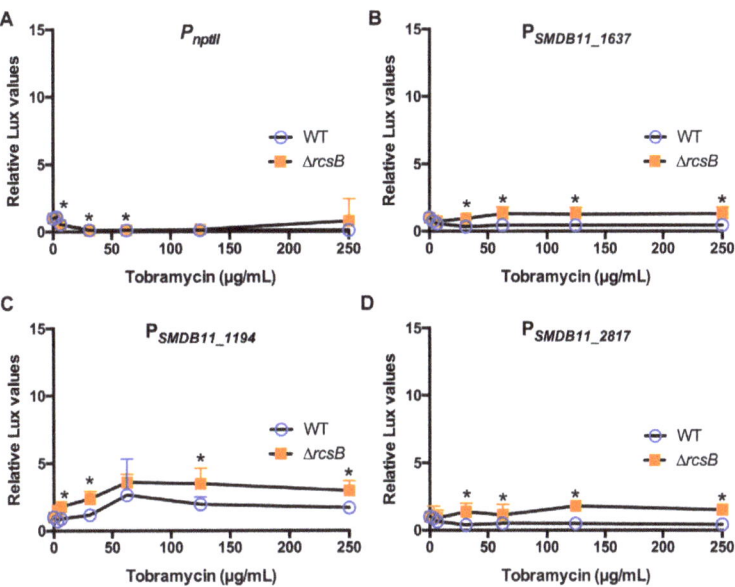

Figure 8. Effect of protein synthesis-targeting antibiotic tobramycin on Rcs-activated promoters (**A–D**). Relative luminescence values were determined by dividing luminescence by optical density after 4 h of antibiotic challenge. The *nptII* promoter (**A**) was unaffected by tobramycin. The experimental promoters were expressed to equal or greater extent in the ΔrcsB mutant than the WT. Mean and standard deviation are shown (*n* = 6–9 are shown). Asterisks (*) indicate statistical differences between groups at the indicated concentrations, $p < 0.05$.

3. Discussion

The major impetus behind this study was to test whether the Rcs system was activated by antibiotics used in topical treatment of keratitis. The results show that several of the antibiotics that are widely used for this purpose indeed do activate the Rcs system. A limitation of the study is that the ocular surface antibiotic pharmacokinetics differ from those in the microplate. While topical antibiotics use very high concentrations, the combined action of blinking and the tears wash away most topical antibiotics in a short time frame. Similarly, antibiotic concentrations reduce over time after application, which could lead to levels that activate the Rcs or other stress response systems. Nevertheless, patients with keratitis are given multiple doses of topical antibiotics each day, and although there are limited studies, data demonstrate measurable quantities of the antibiotics accumulate in the corneal tissue [32,36]. Furthermore, experimental studies with rabbits have shown that concentrations of topically applied antibiotics that mimic clinical treatment regimens are able to kill bacteria in the cornea and even to achieve concentrations sufficient to eliminate bacteria that are considered resistant by systemic standards [39–41]. Therefore, the combination of the highly sensitive promoters and large antibiotic concentration gradients used in this study likely reflects the antibiotic concentrations that bacteria experience during antibiotic therapy for ocular infections.

Additional differences between this in vitro study and the ocular environment include a lack of the innate immune system components that could influence the activity of the antibiotics through synergistic effects or produce envelope stress through other means, such as envelope-targeting defensins and enzymes such as lysozyme and phospholipase A [42,43]. These potential effects will be analyzed in future studies.

Data from this study indicate that the promoters for SMDB11 ORF 1194, 1637, and 2817 are Rcs responsive, given the several $\log_{10}$-fold increase in the Δ*gumB* mutant that required a functional *rcsB* gene. However, it is clear that the selected promoters could also be strongly activated by ciprofloxacin in an Rcs-independent manner. This is not unexpected, as several envelope stress response systems, beyond Rcs, are conserved among the Enterobacterales. For example, in *Salmonella*, the promoter of the *osmB* gene (similar to SMDB11_1687) is activated by both the Rcs and the RpoS stress response systems [28], suggesting that individual stress response genes are controlled by multiple regulatory systems. The use of the Δ*rcsB* strain in addition to the WT enabled clear identification of Rcs-dependent activation of the reporters by ocular antibiotics.

Remarkably, even antibiotics that *S. marcescens* strain K904 was highly resistant to, such as polymyxin B and vancomycin, elicited strong activation of the Rcs system. These results suggest that the antibiotics are still capable of perturbing the envelope, even if they are not able to prevent growth. In general, the three different promoters reacted similarly to each antibiotic, with the notable exception of cefazolin, which only activated the SMDB11_1194 promoter. This may be due to differential promoter elements that make this promoter more sensitive than the others to Rcs function. Polymyxin B has been shown to activate the Rcs system in polymyxin B susceptible *S. enterica* at subinhibitory levels, and this was postulated to be driven by polymyxin B's selective permeabilization of the outer membrane to hydrophobic compounds at low concentrations [11,44]. Several other antibiotics that directly or indirectly affect membrane permeability, including β-lactam, fluoroquinolone, and macrolide antibiotics, are likely capable of the same effect [45].

Of interest, ciprofloxacin appeared to activate some of these promoters to a greater extent in the Δ*rcsB* mutant. This suggests that Rcs may actively inhibit other stress response systems under normal situations. Consistent with this observation, previous studies have demonstrated a complex interplay between the Rcs system and other envelope stress response systems [46–48]. Beyond Rcs, there are other envelope sensing stress response systems in the Enterobacterales, including RpoS, the Cpx system, the phage response proteins, EnvZ/OmpR, and others (reviewed by [46–48]). Very few studies have evaluated the roles of these proteins in *Serratia* species; however, studies have demonstrated pleiotropic roles for Cpx, OmpR, and RpoS in the control of pathogenesis-relevant phenotypes, such as biofilm formation, and secreted enzymes and cytotoxic secondary metabolite production in *Serratia* species [49–52]. The activation of the Rcs system, as noted above, is correlated with changes that drive virulence-associated phenotypes, such as biofilm formation [16,17]. The ability of antibiotics to promote these phenotypes through the Rcs system during ocular infections will be evaluated in subsequent studies.

During the course of this study, another group reported on the production of a Rcs-dependent fluorescent reporter system for *E. coli* [9]. This was subsequently and cleverly used to screen small molecule libraries for activators of the Rcs system, with the concept that the identified molecules may be evaluated and developed as envelope-targeting antimicrobials [53]. Therefore, Rcs-reporter systems can be used for both basic biomedical research and applied studies, and the reporters generated in this study could be useful to a variety of researchers.

4. Materials and Methods
4.1. Bacterial Growth and Media

Bacteria (Table S1) were maintained in glycerol stocks at −80 °C and streaked out on lysogeny broth (LB) agar [54] before use. Single colonies were grown in LB broth with aeration on a tissue culture rotor (New Brunswick Tc-7, New Brunswick, NJ, USA). Gentamicin (10 µg/mL) was used to maintain plasmids. Plasmids were moved into *S. marcescens* by conjugation, and tetracycline (10 µg/mL) was used for selection against donor *E. coli* [55], as previously described. Antibiotics were obtained from Sigma-Aldrich (St. Louis, MO, USA) unless otherwise noted.

4.2. Generation of Luminescence Reporters

The *pigA* promoter on plasmid pMQ713 [56] was replaced with the SMDB11_1194, SMDB11_1637, and SMDB11_2817 using yeast homologous recombination, as previously described [57,58]. Plasmids are listed in Table S1. The pMQ713 plasmid was linearized by restriction enzyme digestion with EcoR1 and Sal1 (New England Biolabs, Ipswich, MA, USA). DNA for the three promoter regions were synthesized as linear double-stranded DNA fragments (Integrated DNA Technologies, Coralville, IA, USA) that include DNA for the promoter region and for site-directed recombination with pMQ713 that places the *luxCDABE* reporter under transcriptional control of the respective promoter (listed in Table S2). The lengths of the cloned promoters were 338 bp for SMDB11_1194, 354 bp for SMDB11_1637, and 337 bp for SMDB11_2817. To generate the *nptII*-driven *luxCDABE* plasmid, the *tdtomato* gene from pMQ414 was digested with BamH1 and EcoR1 enzymes, and the *luxCDABE* operon was amplified by PCR from pMQ670 [59] using primers 3805 and 3806 via PrimeSTAR DNA polymerase (Takara Bio, San Jose, CA, USA). The linearized plasmid and *luxCDABE* amplicon were combined as above. The plasmids were isolated, and the cloned promoter region was sequenced to validate the constructs.

4.3. Luminescent Reporter Assays

Strains of *S. marcescens* bearing luminescent reporter plasmids were taken from a −80 °C freezer and grown on LB agar with tetracycline (10 µg/mL) and gentamicin (10 µg/mL) for 18 h at 30 °C. Single colonies were grown in LB broth with gentamicin in test tubes, which were aerated on a tissue culture rotor for 18–20 h at 30 °C. For reporter verification experiments, the cultures were measured for growth by evaluating optical density at λ = 600 nm (OD_{600}) and luminescence at the 527 nm setting from 200 µL samples in black-sided, clear-bottomed 96-well plates (ThermoFisher, Waltham, MA, USA, product 165305) using plate readers (Molecular Devices SpectraMax M3 and L, San Jose, CA, USA). Relative luminescence units (RLU) values were determined by dividing the raw luminescence values by optical density values.

For antibiotic effect on promoter activity experiments, cultures were normalized by measuring optical density at OD_{600} across a 1-cm path length cuvette with a spectrophotometer (Molecular Devices SpectraMax M3). The assay was conducted in 96-well black-sided, optical bottom plates as above. Two-fold serial dilutions of the antibiotics were performed with a multichannel pipette, and the bacteria were then added to a final concentration of OD_{600} = 0.05 (~9 × 10^7 CFU/mL). The plate was incubated for 4 h at 37 °C in a plastic bag with a dampened paper towel. At 0 and 4 h, luminescence and OD_{600} values were obtained as above. To obtain RLU values, luminescence values were divided by optical density and normalized to RLU values from the no antibiotic challenge control wells.

The antibiotics and maximum concentrations used in this study were polymyxin B at 10 mg/mL (Sigma, St. Louis, MO, USA, product 5291), vancomycin at 5 mg/mL (Fresenius-Kabi, Bad Homburg, Germany, product C22110), cefazolin at 5 mg/mL (WG Critical Care, Paramus, NJ, USA, product 44567-707-25), ceftazidime at 5 mg/mL (Sigma, product C-3809), tobramycin at 1 mg/mL (XGen Pharmaceuticals, Horseheads, NY, USA, product 39882-0412-1), and ciprofloxacin at 0.3 mg/mL (LKT Labs, St. Paul, MN, USA, product C3262). Stock solutions of antibiotic were prepared in a sterile 15-mL polypropylene centrifuge (Corning, Corning, NY, USA) tube by dissolving the solid antibiotics in lysogeny broth (LB). To ensure sterility, the antibiotic solution was filtered using a PVDF 0.22-µm filter (Millipore SLGVR33RB, Cork, Ireland) into a new sterile polypropylene centrifuge tube. All samples were stored at 4 °C when not in use. The antibiotic gentamicin (10 µg/mL) was added into the assay samples for all trials to maintain the plasmids.

4.4. Minimum Inhibitory Concentration Analysis

MIC values were determined by Epsilometer (E-test) testing (bioMérieux Inc., Durham, NC, USA) for cefazolin, ceftazidime, vancomycin, tobramycin, gentamicin, polymyxin B, and ciprofloxacin. In brief, an overnight growth of bacteria was adjusted to a turbidity standard of 0.5 McFarland (~1.2×10^8 CFU/mL) and overlayed with swab streaking on Mueller Hinton agar. E-test strips were placed onto the agar and allowed to incubate for 24 h at 37 °C. The MIC gradients were visually determined and recorded after incubation following the manufacturer's guidelines.

4.5. Statistical Analysis

Tests were performed using Prism software (GraphPad, San Diego, CA, USA). One-way ANOVA with Tukey's post-test was used to compare multiple groups and Student's *t*-test was used to compare between pairs. For this study, *p*-values of less than 0.05 were considered significant.

5. Conclusions

In this study, luminescence reporters for Rcs-stress system activation were generated for use in bacteria of the Enterobacterales order. This stress system induces major transcriptional changes in response envelope stresses that result in increased capsule production and biofilm formation. Using these reporters, the Rcs response of the ocular pathogen *S. marcescens* to antibiotics used for the treatment of keratitis was evaluated. Several classes of antibiotics used to treat keratitis induced the Rcs system even when the test bacterium was highly resistant to the respective antibiotic. These data suggest that topical treatment of ocular infections with antibiotics may lead to Rcs-dependent phenotypic changes that aid in bacterial antibiotic tolerance.

Supplementary Materials: The following are available online at https://www.mdpi.com/article/10.3390/antibiotics10091033/s1: Table S1: *S. marcescens* strains and plasmids used in this study; Table S2: Nucleic acids used in this study; Figure S1: Diagram of pMQ747 used in this study.

Author Contributions: Conceptualization: R.M.Q.S.; methodology: R.M.Q.S., R.P.K. and N.S.H.; investigation: N.S.H., N.A.S., K.M.L., E.G.R. and R.P.K.; data curation: E.G.R., R.P.K. and R.M.Q.S.; writing—original draft preparation: N.S.H., R.P.K. and R.M.Q.S.; writing—review and editing: N.S.H., N.A.S., E.G.R., R.P.K. and R.M.Q.S.; funding acquisition: R.M.Q.S. All authors have read and agreed to the published version of the manuscript.

Funding: This work was supported by Research to Prevent Blindness (unrestricted funds), the Eye and Ear Foundation of Pittsburgh, and National Institute of Health grants P30EY08098 (to Department of Ophthalmology) and R01EY027331 (to R.M.Q.S.).

Institutional Review Board Statement: Not applicable.

Conflicts of Interest: The authors declare no conflict of interest.

References

1. Alexandrakis, G.; Alfonso, E.C.; Miller, D. Shifting trends in bacterial keratitis in south Florida and emerging resistance to fluoroquinolones. *Ophthalmology* **2000**, *107*, 1497–1502. [CrossRef]
2. Hume, E.B.; Willcox, M.D. Emergence of *Serratia marcescens* as an ocular surface pathogen. *Arch. Soc. Esp. Oftalmol.* **2004**, *79*, 475–477.
3. Mah-Sadorra, J.H.; Najjar, D.M.; Rapuano, C.J.; Laibson, P.R.; Cohen, E.J. *Serratia* corneal ulcers: A retrospective clinical study. *Cornea* **2005**, *24*, 793–800. [CrossRef] [PubMed]
4. Mahlen, S.D. Serratia infections: From military experiments to current practice. *Clin. Microbiol. Rev.* **2011**, *24*, 755–791. [CrossRef] [PubMed]
5. Richards, M.J.; Edwards, J.R.; Culver, D.H.; Gaynes, R.P. Nosocomial infections in combined medical-surgical intensive care units in the United States. *Infect. Control Hosp. Epidemiol.* **2000**, *21*, 510–515. [CrossRef] [PubMed]
6. Stock, I.; Grueger, T.; Wiedemann, B. Natural antibiotic susceptibility of strains of *Serratia marcescens* and the *S. liquefaciens* complex: *S. liquefaciens* sensu stricto, *S. proteamaculans* and *S. grimesii*. *Int. J. Antimicrob. Agents* **2003**, *22*, 35–47. [CrossRef]
7. Kowalski, R.P.; Kowalski, T.A.; Shanks, R.M.; Romanowski, E.G.; Karenchak, L.M.; Mah, F.S. In vitro comparison of combination and monotherapy for the empiric and optimal coverage of bacterial keratitis based on incidence of infection. *Cornea* **2013**, *32*, 830–834. [CrossRef]
8. Wall, E.; Majdalani, N.; Gottesman, S. The Complex Rcs Regulatory Cascade. *Annu. Rev. Microbiol.* **2018**, *72*, 111–139. [CrossRef]

9. Steenhuis, M.; Ten Hagen-Jongman, C.M.; van Ulsen, P.; Luirink, J. Stress-based high-througput screen assays to identify inhibitors of cell envelope biogenesis. *Antibiotics* **2020**, *9*, 808. [CrossRef] [PubMed]
10. Erickson, K.D.; Detweiler, C.S. The Rcs phosphorelay system is specific to enteric pathogens/commensals and activates *ydeI*, a gene important for persistent *Salmonella* infection of mice. *Mol. Microbiol.* **2006**, *62*, 883–894. [CrossRef] [PubMed]
11. Farris, C.; Sanowar, S.; Bader, M.W.; Pfuetzner, R.; Miller, S.I. Antimicrobial peptides activate the Rcs regulon through the outer membrane lipoprotein RcsF. *J. Bacteriol.* **2010**, *192*, 4894–4903. [CrossRef]
12. Laubacher, M.E.; Ades, S.E. The Rcs phosphorelay is a cell envelope stress response activated by peptidoglycan stress and contributes to intrinsic antibiotic resistance. *J. Bacteriol.* **2008**, *190*, 2065–2074. [CrossRef] [PubMed]
13. Konovalova, A.; Mitchell, A.M.; Silhavy, T.J. A lipoprotein/b-barrel complex monitors lipopolysaccharide integrity transducing information across the outer membrane. *Elife* **2016**, *5*, e15276. [CrossRef] [PubMed]
14. Hirakawa, H.; Nishino, K.; Yamada, J.; Hirata, T.; Yamaguchi, A. Beta-lactam resistance modulated by the overexpression of response regulators of two-component signal transduction systems in *Escherichia coli*. *J. Antimicrob. Chemother.* **2003**, *52*, 576–582. [CrossRef] [PubMed]
15. Lewis, K. Multidrug tolerance of biofilms and persister cells. *Curr. Top. Microbiol. Immunol.* **2008**, *322*, 107–131. [CrossRef]
16. Stella, N.A.; Brothers, K.M.; Callaghan, J.D.; Passerini, A.M.; Sigindere, C.; Hill, P.J.; Liu, X.; Wozniak, D.J.; Shanks, R.M.Q. An IgaA/UmoB-family protein from *Serratia marcescens* regulates motility, capsular polysaccharide, and secondary metabolite production. *Appl. Environ. Microbiol.* **2018**, *84*, e02517–e02575. [CrossRef] [PubMed]
17. Clarke, D.J. The Rcs phosphorelay: More than just a two-component pathway. *Future Microbiol.* **2010**, *5*, 1173–1184. [CrossRef]
18. Smith, L.M.; Jackson, S.A.; Malone, L.M.; Ussher, J.E.; Gardner, P.P.; Fineran, P.F. The Rcs stress response inversely controls surface and CRISPR-Cas adaptive immunity to discriminate plasmids and phages. *Nat. Microbiol.* **2021**, *6*, 162–172. [CrossRef]
19. Brothers, K.M.; Callaghan, J.D.; Stella, N.A.; Bachinsky, J.M.; AlHigaylan, M.; Lehner, K.L.; Franks, J.M.; Lathrop, K.L.; Collins, E.; Schmitt, D.M.; et al. Blowing epithelial cell bubbles with GumB: ShlA-family pore-forming toxins induce blebbing and rapid cellular death in corneal epithelial cells. *PLoS Pathog.* **2019**, *15*, e1007825. [CrossRef] [PubMed]
20. Di Venanzio, G.; Stepanenko, T.M.; Garcia Vescovi, E. *Serratia marcescens* ShlA pore-forming toxin is responsible for early induction of autophagy in host cells and is transcriptionally regulated by RcsB. *Infect. Immun.* **2014**, *82*, 3542–3554. [CrossRef]
21. Pan, X.; Tang, M.; You, J.; Liu, F.; Sun, C.; Osire, T.; Fu, W.; Yi, G.; Yang, T.; Yang, S.T.; et al. Regualtor RcsB controls prodigiosin synthesis and various cellular processes in *Serratia marcescens* JNB5-1. *Appl. Environ. Microbiol.* **2020**, in press. [CrossRef]
22. Romanowski, E.G.; Stella, N.A.; Romanowski, J.E.; Yates, K.A.; Dhaliwal, D.K.; St Leger, A.J.; Shanks, R.M.Q. The Rcs stress response system regulator GumB modulates *Serratia marcescens* induced inflammation and bacterial proliferation in a rabbit keratitis model and cytotoxicity in vitro. *Infect. Immun.* **2021**, *89*, e00111–e00121. [CrossRef]
23. Lin, A.; Rhee, M.K.; Akpek, E.K.; Amescua, G.; Farid, M.; Garcia-Ferrer, F.J.; Varu, D.M.; Musch, D.C.; Dunn, S.P.; Mah, F.S.; et al. Bacterial Keratitis Preferred Practice Pattern(R). *Ophthalmology* **2019**, *126*, P1–P55. [CrossRef]
24. Wang, D.; Qi, M.; Calla, B.; Korban, S.S.; Clough, S.J.; Cock, P.J.; Sundin, G.W.; Toth, I.; Zhao, Y. Genome-wide identification of genes regulated by the Rcs phosphorelay system in *Erwinia amylovora*. *Mol. Plant-Microbe Interact.* **2012**, *25*, 6–17. [CrossRef]
25. Bury-Moné, S.; Nomane, Y.; Reymond, N.; Barbet, R.; Jacquet, E.; Imbeaud, S.; Jacq, A.; Bouloc, P. Global analysis of extracytoplasmic stress signaling in *Escherichia coli*. *PLoS Genet.* **2009**, *5*, e1000651. [CrossRef] [PubMed]
26. Boulanger, A.; Frances-Charlot, A.; Conter, A.; Castanié-Cornet, M.-P.; Cam, K.; Gutierrez, C. Multistress regulation in *Escherichia coli*: Expression of *osmB* involves two independent promoters responding either to sigmaS or to the RcsCDB His-Asp phosphorelay. *J. Bacteriol.* **2005**, *187*, 3282–3286. [CrossRef] [PubMed]
27. Howery, K.E.; Clemmer, K.M.; Rather, P.N. The Rcs regulon in *Proteus mirabilis*: Implications for motility, biofilm formation, and virulence. *Curr. Genet.* **2016**, *62*, 775–789. [CrossRef]
28. Huesa, J.; Giner-Lamia, J.; Pucciarelli, M.G.; Paredes-Martínez, F.; García del Portillo, F.; Marina, A.; Casino, P. Structure-based analysis of *Salmonella* RcsB variants unravel new features of the Rcs regulon. *Nucleic Acids Res.* **2021**, *49*, 2357–2374. [CrossRef]
29. Hinchliffe, S.J.; Howard, S.L.; Huang, Y.H.; Clarke, D.J.; Wren, B.W. The importance of the Rcs phosphorelay in the survival and pathogenesis of the enteropathogenic yersiniae. *Microbiology* **2008**, *154*, 1117–1131. [CrossRef]
30. Lehner, K.M.; Stella, N.A.; Calvario, R.C.; Shanks, R.M.Q. mCloverBlaster: A tool to make markerless deletions and fusion using lambda red and I-SceI in Gram-negative bacterial genomes. *J. Microbiol. Methods* **2020**, *178*. [CrossRef]
31. Cunha, B.A. *Antibiotic Essentials*; Physicians Press: Royal Oak, MI, USA, 2002.
32. Healy, D.P.; Holland, E.J.; Nordlund, M.L.; Dunn, S.; Chow, C.; Lindstrom, R.L.; Hardten, D.; Davis, E. Concentrations of levofloxacin, ofloxacin, and ciprofloxacin in human corneal stromal tissue and aqueous humor after topical administration. *Cornea* **2004**, *23*, 255–263. [CrossRef]
33. Tajima, K.; Miyake, T.; Koike, N.; Hattori, T.; Kumakura, S.; Yamaguchi, T.; Matsumoto, T.; Fujita, K.; Kuroda, M.; Ito, N.; et al. In vivo challenging of polymyxins and levofloxacin eye drops against multidrug-resistant *Pseudomonas aeruginosa* keratits. *J. Infect. Chemother.* **2014**, *20*, 343–349. [CrossRef]
34. Protzko, E.; Bowman, L.; Abelson, M.; Shapiro, A. Phase 3 safety comparisons for 1.0% azithromycin in polymeric mucoadhesive eye drops versus 0.3% tobramycin eye drops for bacterial conjunctivitis. *Investig. Ophthalmol. Vis. Sci.* **2007**, *48*, 3425–3429. [CrossRef]

35. Romanowski, E.G.; Romanowski, J.E.; Shanks, R.M.Q.; Yates, K.A.; Mammen, A.; Dhaliwal, D.K.; Jhanji, V.; Kowalski, R.P. Topical vancomycin 5% is more efficacious than 2.5% and 1.25% for reducing viable methicillin-resistant *Staphylococcus aureus* in infectious keratitis. *Cornea* **2020**, *39*, 250–253. [CrossRef]
36. Cahane, M.; Ben Simon, G.J.; Barequet, I.S.; Grinbaum, A.; Diamanstein-Weiss, L.; Goller, O.; Rubinstein, E.; Avni, I. Human corneal stromal tissue concentration after consecutive doses of topically applied 3.3% vancomycin. *Br. J. Ophthalmol.* **2004**, *88*, 22–24. [CrossRef] [PubMed]
37. CLSI. *Performance Standards for Antimicrobial Susceptibility Testing 24th Informational Supplement*; CLSI Document M100-S24; Clinical and Laboratory Standards Institute: Wayne, PA, USA, 2014; pp. 51–57.
38. Lin, Q.Y.; Tsai, Y.L.; Liu, M.C.; Lin, W.C.; Hsueh, P.R.; Liaw, S.J. *Serratia marcescens arn*, a PhoP-regulated locus necessary for polymyxin B resistance. *Antimicrob. Agents Chemother.* **2014**, *58*, 5180–5190. [CrossRef]
39. Romanowski, E.G.; Mah, F.S.; Yates, K.A.; Kowalski, R.P.; Gordon, Y.J. The successful treatment of gatifloxacin-resistant *Staphylococcus aureus* keratitis with Zymar (gatifloxacin 0.3%) in a NZW rabbit model. *Am. J. Ophthalmol.* **2005**, *139*, 867–877. [CrossRef]
40. Kowalski, R.P.; Romanowski, E.G.; Mah, F.S.; Shanks, R.M.; Gordon, Y.J. Topical levofloxacin 1.5% overcomes in vitro resistance in rabbit keratitis models. *Acta Ophthalmol.* **2010**, *88*, e120–e125. [CrossRef] [PubMed]
41. Kowalski, R.P.; Romanowski, E.G.; Yates, K.A.; Romanowski, J.E.; Grewal, A.; Bilonick, R.A. Is there a role for topical penicillin treatment of *Staphylococcus aureus* keratitis based on elevated corneal concentrations? *J. Clin. Ophthalmol. Optom.* **2018**, *2*, 103.
42. McDermott, A.M. Antimicrobial compounds in tears. *Exp. Eye Res.* **2013**, *117*, 53–61. [CrossRef] [PubMed]
43. Pearlman, E.; Sun, Y.; Roy, S.; Karmakar, M.; Hise, A.G.; Szczotka-Flynn, L.; Ghannoum, M.; Chinnery, H.R.; McMenamin, P.G.; Rietsch, A. Host defense at the ocular surface. *Int. Rev. Immunol.* **2013**, *32*, 4–18. [CrossRef]
44. Vasilchenko, A.S.; Rogozhin, E.A. Sub-inhibitory effects of antimicrobial peptides. *Front. Microbiol.* **2019**, *10*, 1160. [CrossRef] [PubMed]
45. Poole, K. Bacterial stress responses as determinants of antimicrobial resistance. *J. Antimicrob. Chemother.* **2012**, *67*, 2069–2089. [CrossRef]
46. Flores-Kim, J.; Darwin, A.J. Regulation of bacterial virulence gene expression by cell envelope stress responses. *Virulence* **2014**, *5*, 835–851. [CrossRef]
47. Laloux, G.; Collet, J.F. Major Tom to ground control: How lipoproteins communicate extracytoplasmic stress to the decision center of the cell. *J. Bact.* **2017**, *199*, 00216–00217. [CrossRef] [PubMed]
48. Macritchie, D.M.; Raivio, T.L. Envelope Stress Responses. *EcoSal Plus* **2009**, *3*. [CrossRef] [PubMed]
49. Bruna, R.E.; Molino, M.V.; Lazzaro, M.; Mariscotti, J.F.; Garcia Véscovi, E. CpxR-dependent thermoregulation of *Serratia marcescens* PrtA metalloprotease expression and its contribution to bacterial biofilm formation. *J. Bacteriol.* **2018**, *200*, e00006–e00018. [CrossRef] [PubMed]
50. Qin, H.; Liu, Y.; Cao, X.; Jiang, J.; Lian, W.; Qiao, D.; Xu, H.; Cao, Y. RpoS is a peiotropic regulator of motility, biofilm formation, exoenzymes, siderophore and prodigiosin production, and trade-off during prolonged stationary phase in *Serratia marcescens*. *PLoS ONE* **2020**, *15*, e0232549. [CrossRef]
51. Sun, Y.; Wang, L.; Pan, X.; Osire, T.; Fang, H.; Zhang, H.; Yang, S.T.; Yang, T.; Rao, Z. Improved prodigiosin production by relieving CpxR temperature-sensitive inhibition. *Front. Bioeng. Biotechnol.* **2020**, *8*, 344. [CrossRef]
52. Wilf, N.M.; Salmond, G.P. The stationary phase sigma factor, RpoS, regulates the production of a carbapenem antibiotic, a bioactive prodigiosin and virulence in the enterobacterial pathogen *Serratia* sp. ATCC 39006. *Microbiology* **2012**, *158*, 648–658. [CrossRef]
53. Steenhuis, M.; Corona, F.; Ten Hagen-Jongman, C.M.; Vollmer, W.; Lambin, D.; Selhorst, P.; Klaassen, H.; Versele, M.; Chatltin, P.; Luirink, J. Combining cell envelope stress reporter assays in a screening approach to identify BAM complex inhibitors. *ACS Infect. Dis.* **2021**, in press. [CrossRef] [PubMed]
54. Bertani, G. Studies on lysogenesis. I. The mode of phage liberation by lysogenic *Escherichia coli*. *J. Bacteriol.* **1951**, *62*, 293–300. [CrossRef] [PubMed]
55. Shanks, R.M.; Stella, N.A.; Kalivoda, E.J.; Doe, M.R.; O'Dee, D.M.; Lathrop, K.L.; Guo, F.L.; Nau, G.J. A *Serratia marcescens* OxyR homolog mediates surface attachment and biofilm formation. *J. Bacteriol.* **2007**, *189*, 7262–7272. [CrossRef] [PubMed]
56. Romanowski, E.G.; Lehner, K.M.; Martin, N.C.; Patel, K.R.; Callaghan, J.D.; Stella, N.A.; Shanks, R.M.Q. Thermoregulation of prodigiosin biosynthesis by *Serratia marcescens* is controlled at the transcriptional level and requires HexS. *Pol. J. Microbiol.* **2019**, *68*, 43–50. [CrossRef]
57. Shanks, R.M.; Caiazza, N.C.; Hinsa, S.M.; Toutain, C.M.; O'Toole, G.A. *Saccharomyces cerevisiae*-based molecular tool kit for manipulation of genes from gram-negative bacteria. *Appl. Environ. Microbiol.* **2006**, *72*, 5027–5036. [CrossRef]
58. Shanks, R.M.; Kadouri, D.E.; MacEachran, D.P.; O'Toole, G.A. New yeast recombineering tools for bacteria. *Plasmid* **2009**, *62*, 88–97. [CrossRef]
59. Callaghan, J.D.; Stella, N.A.; Lehner, K.M.; Treat, B.R.; Brothers, K.M.; St Leger, A.J.; Shanks, R.M.Q. Generation of Xylose-inducible promoter tools for *Pseudomonas* species and their use in implicating a role for the Type II secretion system protein XcpQ in inhibition of corneal epithelial wound closure. *Appl. Environ. Microbiol.* **2020**, *86*, e00250-20. [CrossRef] [PubMed]

Article

Transcription Factor EepR Is Required for *Serratia marcescens* Host Proinflammatory Response by Corneal Epithelial Cells

Kimberly M. Brothers, Stephen A. K. Harvey and Robert M. Q. Shanks *

Charles T. Campbell Ophthalmic Microbiology Laboratory, Department of Ophthalmology, University of Pittsburgh School of Medicine, Pittsburgh, PA 15213, USA; kmb227@pitt.edu (K.M.B.); stephenaharvey@cs.com (S.A.K.H.)
* Correspondence: shanksrm@upmc.edu; Tel.: +1-412-647-3537

Abstract: Relatively little is known about how the corneal epithelium responds to vision-threatening bacteria from the Enterobacterales order. This study investigates the impact of *Serratia marcescens* on corneal epithelial cell host responses. We also investigate the role of a bacterial transcription factor EepR, which is a positive regulator of *S. marcescens* secretion of cytotoxic proteases and a hemolytic surfactant. We treated transcriptomic and metabolomic analysis of human corneal limbal epithelial cells with wild-type bacterial secretomes. Our results show increased expression of proinflammatory and lipid signaling molecules, while this is greatly altered in *eepR* mutant-treated corneal cells. Together, these data support the model that the *S. marcescens* transcription factor EepR is a key regulator of host-pathogen interactions, and is necessary to induce proinflammatory chemokines, cytokines, and lipids.

Keywords: bacterial infection; *Serratia marcescens*; transcription factor; keratitis; ocular surface; epithelium; cornea; metabolomics

Citation: Brothers, K.M.; Harvey, S.A.K.; Shanks, R.M.Q. Transcription Factor EepR Is Required for *Serratia marcescens* Host Proinflammatory Response by Corneal Epithelial Cells. *Antibiotics* **2021**, *10*, 770. https://doi.org/10.3390/antibiotics10070770

Academic Editors: Mark Willcox, Fiona Stapleton and Debarun Dutta

Received: 21 May 2021
Accepted: 22 June 2021
Published: 24 June 2021

Publisher's Note: MDPI stays neutral with regard to jurisdictional claims in published maps and institutional affiliations.

Copyright: © 2021 by the authors. Licensee MDPI, Basel, Switzerland. This article is an open access article distributed under the terms and conditions of the Creative Commons Attribution (CC BY) license (https://creativecommons.org/licenses/by/4.0/).

1. Introduction

The cornea, the transparent, anterior layer of the eye, is essential for vision and protected by numerous host immune factors, the tear film [1,2], and the corneal epithelium [3,4]. When the epithelium is damaged or compromised, it permits entry of microbes into the stroma where they can multiply and cause damage to the ocular tissues; the progression of infection is rapid, sometimes leading to corneal perforation from bacterial proteases and from the ensuing inflammatory response [5–9].

Serratia marcescens is a gram negative pathogen from the order Enterobacterales frequently isolated from contact lenses, and associated with ocular infections [10–12]. Bacteria are linked with chronic infections, non-healing wounds, and are thought to prevent wound closure; however, the impact of bacteria on corneal infection and wound healing is poorly understood [13–15]. Our previous study identified *S. marcescens* LPS as being sufficient to inhibit corneal epithelial wound closure and further identified transposon insertions in genes that rendered the bacterium unable to inhibit corneal cell migration, but the role of these genes in ocular surface host-pathogen interactions was not characterized [16]. One mutation mapped to the *eepR-eepS* locus, that codes for a hybrid two-component transcription factor system involved in virulence factor secretion, cytotoxicity to mammalian cells, and proliferation in a rabbit keratitis model [17–19].

Previous studies have evaluated the impact of bacteria on the global transcriptomic response of corneal cells, but this has only been done with *Pseudomonas aeruginosa* and *Staphylococcus aureus* [20–23]. In this study, the role of the EepR transcriptional regulator in the corneal epithelial cell transcriptional and small molecule response to *S. marcescens* was evaluated. We report that in contrast to other pathogens, mutation of one bacterial transcription factor in *S. marcescens* had a broad impact on epithelial cell responses, including reduced expression of inflammatory markers and lipid metabolism genes.

2. Results

2.1. HCLE Cells Exposed to eepR Mutant S. marcescens Secretomes Have an Attenuated Inflammatory Response Compared with Wild-type Treated HCLE Cells

To increase our understanding of the corneal response to an order of bacteria not previously tested, a global transcriptional analysis of the HCLE cells was performed. Here we used a wild-type (WT), low cytotoxicity [24] isolate of *S. marcescens* (PIC3611), and an isogenic strain with a deletion in the *eepR* gene that was previously described [19] to further investigate EepR's role in how bacteria influence corneal biology. In this study, bacterial secretomes were used to stimulate corneal cells because we have previously shown wild-type secretomes to strongly influence the behavior of a human corneal epithelial cell line and because secretomes are less toxic to corneal cells [16,25,26]. Confluent monolayers of the human corneal limbal epithelial (HCLE) cell line were first exposed to *S. marcescens* WT secretome for 0, 1, 2, 3, 4, and 5 h to determine the time frame for maximal stimulation by assessing levels of the cytokine TNFα. The 5 h exposure time point was chosen based upon our preliminary findings (data not shown) and from a previous ELISA-based study of human corneal epithelial cell inflammatory response to *S. marcescens* [27].

Next, we compared the transcriptomes of mock-treated (LB medium in equal volume as secretomes) corneal cells with those exposed to normalized secretomes from WT or *eepR* cells. Lower case *eepR* refers to the mutant strain. As noted in Materials and Methods, 21,932 microarray panels (unique target sequences) yielded reliable data; valid changes between WT secretome-treated and mock-treated cells occurred in only 2510 panels (11.4%), and of those, only 915 (4.2%) were modulated by 2-fold or more (examples in Tables 1 and 2). In contrast, valid changes between *eepR* secretome-treated and mock-treated cells occurred in only 798 panels (3.6%), and of those, only 241 (1.1%) were modulated by 2-fold or more (examples in Tables 3 and 4). Over half of the *eepR* secretome-modulated panels (138, 57%) were present in the WT-treatment group also (see nine genes in common between Tables 1 and 3, eight genes in common between Tables 2 and 4), and the direction of modulation was concordant between treatments for all these panels except SPRY2, which was increased by WT treatment and decreased by *eepR*. Visual inspection showed that within this group of 138 genes, whatever the direction of change caused by *eepR* (increase or decrease), its magnitude was always less than that caused by WT. However, some genes outside this group showed greater modulation by *eepR* than by WT. Accordingly, the scaled *eepR* response ($eepR$ − control)/ | (WT − control) | was also calculated (Tables 5 and 6).

The 915 panels modulated by WT were submitted to Ingenuity Pathway Analysis software (Qiagen, Germantown, MD, USA), yielding 24 significantly enriched ($p < 0.05$) canonical pathways which had adequate z-scores (| z | > 2; see Table 7). At least nine of these pathways address direct or indirect immune functions. When submitted for analysis separately, the 798 *eepR* modulated panels only yielded three significantly enriched pathways, two of which were also WT-modulated (see Table 7). The third pathway (GNRH Signaling) was not significantly enriched by WT treatment. In *S. marcescens* WT secretome-treated HCLEs versus mock-treated cells, the twenty-five most upregulated genes (9.1- to 56.6-fold increase) included genes involved in inflammatory signaling pathways (Table 1). Genes with the greatest decrease (4.9- to 50-fold decrease) in WT secretome-treated HCLEs were those involved in nucleosome assembly, phospholipid metabolic processes, and transcription (Table 2). Moreover, HCLEs-treated with *eepR* secretome showed decreased upregulation of genes for proinflammatory factors; however, genes involved in cell to cell adhesion, leukocyte chemotaxis, transport, and signaling were upregulated (Table 3). Genes with the greatest decrease in *eepR versus* mock-treated secretomes were those involved in nucleic acid binding, transport, and transcription (Table 4).

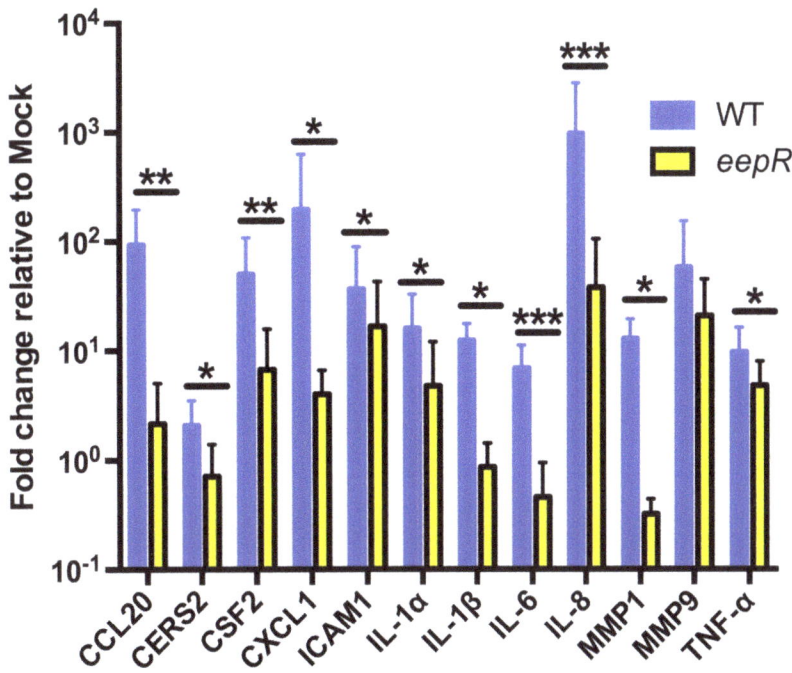

Figure 1. qRT-PCR of pathway markers confirmed microarray analysis. Graph represents the fold change in gene expression relative to mock (LB) treatment. HCLE cells were exposed to LB, WT, and *eepR* transcriptomes of 5 h. Gene expression was normalized to GAPDH expression. Means (n = 4–8, n = 3 for IL-1α) and SD are shown. ΔΔCT values were compared by ANOVA with Bonferroni's post-test, one asterisk (*) indicate $p < 0.05$, two indicate (**) $p < 0.01$, and three (***) indicate $p < 0.001$.

Table 1. Twenty-five genes with the greatest expression increase in cells treated with WT vs. mock secretomes.

Gene symbol	Entrez Gene number	Mean of Normalized Expression, Duplicates			Expression Ratios				Scaled eepR	Biological Function
		LB control	WT *Serratia*	*eepR* mutant	**WT/cont**	*eepR*/cont	*eepR*/WT	WT/*eepR*	$\frac{(eppR-cont)}{(WT-cont)}$	
CXCL8 *	3576	41	2346	145	**56.6**	3.5	0.1	16.2	0.04	Inflammatory cytokine
CXCL1 *	2919	51	1452	566	**28.4**	11.1	0.4	2.6	0.37	Inflammatory cytokine
CCL20 *	6364	297	8224	1757	**27.7**	5.9	0.2	4.7	0.18	Inflammatory cytokine
ITGB8	3696	4	91	38	**25.7**	10.8	0.4	2.4	0.40	Integrin-mediated cell adhesion
CXCL3	2921	35	830	108	**23.7**	3.1	0.1	7.7	0.09	Chemotaxis
GFPT2	9945	5	96	39	**18.4**	7.4	0.4	2.5	0.37	Glutamine fructose-6-phosphate transaminase
CSF2 *	1437	74	1311	192	**17.7**	2.6	0.1	6.8	0.10	granulocyte macrophage colony-stimulating factor receptor binding
LIF	3976	71	1164	93	**16.4**	1.3	0.1	12.5	0.02	TGF Beta Signaling
CSF3	1440	47	746	146	**16.0**	3.1	0.2	5.1	0.14	granulocyte colony-stimulating factor receptor binding

Table 1. Cont.

		Mean of Normalized Expression, Duplicates			Expression Ratios				Scaled eepR	
MMP1 *	4312	104	1639	114	**15.7**	1.1	0.1	14.4	0.01	Proteolysis
CXCL2	2920	83	1163	428	**14.1**	5.2	0.4	2.7	0.32	Chemokine
MTSS1	9788	11	137	31	**12.9**	2.9	0.2	4.4	0.16	Actin binding
HCAR3	8843	306	3840	589	**12.6**	1.9	0.2	6.5	0.08	G-protein coupled receptor signaling
IL20	50604	31	379	26	**12.3**	0.8	0.1	14.6	−0.01	Receptor binding
TNFAIP2	7127	20	236	198	**11.7**	9.9	0.8	1.2	0.83	Angiogenesis
ICAM1 *	3383	40	445	217	**11.3**	5.5	0.5	2.1	0.44	T cell antigen processing and presentation
IL36G	56300	92	1011	536	**11.0**	5.8	0.5	1.9	0.48	Positive regulation of cytokine production
SQSTM1	8878	8	81	28	**10.7**	3.7	0.3	2.9	0.28	Positive regulation of protein phosphorylation
MMP10	4319	100	1064	92	**10.7**	0.9	0.1	11.6	−0.01	Proteolysis
PRDM1	639	139	1394	146	**10.1**	1.1	0.1	9.5	0.01	Negative regulation of transcription from RNA polymerase II
TRAF1	7185	16	160	54	**10.0**	3.4	0.3	3.0	0.27	Apoptosis
IL1R2	7850	66	640	412	**9.7**	6.2	0.6	1.6	0.60	Immune response
IL24	11009	474	4413	678	**9.3**	1.4	0.2	6.5	0.05	Apoptosis
MMP9 *	4318	408	3792	2017	**9.3**	4.9	0.5	1.9	0.48	Proteolysis
IL6 *	3569	60	545	152	**9.1**	2.5	0.3	3.6	0.19	Inflammatory cytokine

Seven genes PCR verified (*), see Figure 1. Nine genes in bold also appear in Table 3: "Greatest expression increase in cells treated with *eepR* vs. mock secretomes".

Table 2. Twenty-five genes with the greatest expression decrease in cells treated with WT vs. mock secretomes.

		Mean of Normalized Expression, Duplicates			Expression Ratios				Scaled eepR	
Gene Symbol	Entrez Gene number	LB control	WT *Serratia*	eepR mutant	**WT/cont**	eepR/cont	eepR/WT	WT/eepR	$\frac{(eppR-cont)}{(WT-cont)}$	Biological Function
TXNIP	10628	2659	64	320	**0.02**	0.1	5.0	0.2	−0.90	Negative regulation of transcription from RNA polymerase II
CTGF	1490	1245	55	35	**0.04**	0.0	0.6	1.6	−1.02	Cartilage condensation
236865_at	—	117	7	24	**0.06**	0.2	3.5	0.3	−0.84	Unknown
ARRDC4	91947	1338	96	285	**0.07**	0.2	3.0	0.3	−0.85	Positive regulation of ubiquitin-protein ligase activity
LOC100287896	100287896	81	6	38	**0.08**	0.5	5.9	0.2	−0.57	Unknown
NAP1L3	4675	35	4	31	**0.10**	0.9	8.6	0.1	−0.14	Nucleosome assembly
RP4-813F11.4	—	146	19	13	**0.13**	0.1	0.7	1.5	−1.05	Unknown
HJURP	55355	747	105	430	**0.14**	0.6	4.1	0.2	−0.49	Nucleosome assembly
PIK3R3	8503	95	14	70	**0.14**	0.7	5.2	0.2	−0.31	Phospholipid metabolic process
SLC26A7	115111	24	4	5	**0.14**	0.2	1.3	0.8	−0.95	Gastric acid secretion
ARRDC3	57561	257	40	113	**0.15**	0.4	2.9	0.4	−0.66	Temperature homeostasis
ZNF750	79755	148	24	62	**0.16**	0.4	2.6	0.4	−0.69	Transcription, DNA-dependent

Table 2. Cont.

		Mean of Normalized Expression, Duplicates			Expression Ratios				Scaled eepR	
GPX8	493869	92	15	86	0.16	0.9	5.8	0.2	−0.08	Response to oxidative stress
MECOM	2122	154	25	65	0.16	0.4	2.6	0.4	−0.69	Neutrophil homeostasis
ENC1	8507	379	64	169	0.17	0.4	2.6	0.4	−0.67	Multicellular organismal development
THAP2	83591	88	15	11	0.17	0.1	0.7	1.4	−1.05	Nucleic acid binding
1560973_a_at	—	34	6	16	0.18	0.5	2.7	0.4	−0.63	Unknown
ZNF658	26149	56	10	56	0.19	1.0	5.4	0.2	−0.00	Transcription, DNA-dependent
ST6GALNAC5	81849	76	14	44	0.19	0.6	3.1	0.3	−0.51	Protein glycosylation
AOC3	8639	84	16	8	0.19	0.1	0.5	2.0	−1.11	Cell adhesion
AKNAD1	254268	67	13	23	0.20	0.3	1.7	0.6	−0.82	Cytoplasm
FAM83D	81610	1588	313	1235	0.20	0.8	3.9	0.3	−0.28	Cell cycle
242708_at	—	44	9	8	0.20	0.2	0.9	1.1	−1.01	Unknown
ZC3H6	376940	99	20	32	0.21	0.3	1.6	0.6	−0.85	Nucleic acid binding
* FAM72A	554282	1976	413	1063	0.21	0.5	2.6	0.4	−0.58	Cytoplasm

* Full designation of bottom row: FAM72A /// FAM72B /// FAM72C /// FAM72D: Entrez numbers 554282 /// 653820 /// 728833 /// 729533. Eight genes in bold also appear in Table 4: "Greatest expression decrease in cells treated with *eepR* vs. mock secretomes".

Table 3. Twenty-five genes with the greatest expression increase in cells treated with *eepR* vs. mock secretomes.

		Mean of Normalized Expression, Duplicates			Expression Ratios				Scaled eepR	
Gene Symbol	Entrez Gene number	LB control	WT *Serratia*	*eepR* mutant	WT/cont	*eepR*/cont	*eepR*/WT	WT/*eepR*	$\frac{(eepR-cont)}{(WT-cont)}$	Biological Function
CXCL1	2919	51	1452	566	28.4	**11.1**	0.4	2.6	0.37	Inflammatory cytokine
ITGB8	3696	4	91	38	25.7	**10.8**	0.4	2.4	0.40	Integrin-mediated cell adhesion
TNFAIP2	7127	20	236	198	11.7	**9.9**	0.8	1.2	0.83	Angiogenesis
OLR1	4973	155	1302	1195	8.4	**7.7**	0.9	1.1	0.91	Proteolysis
IL1R2	7850	66	640	412	9.7	**6.2**	0.6	1.6	0.60	Immune response
CCL20	6364	297	8224	1757	27.7	**5.9**	0.2	4.7	0.18	Inflammatory cytokine
IL36G	56300	92	1011	536	11.0	**5.8**	0.5	1.9	0.48	Positive regulation of cytokine production
SLC2A6	11182	34	97	189	2.9	**5.6**	1.9	0.5	2.45	Transport
ICAM1	3383	40	445	217	11.3	**5.5**	0.5	2.1	0.44	T cell antigen processing and presentation
CXCL2	2920	83	1163	428	14.1	**5.2**	0.4	2.7	0.32	Chemokine
MMP9	4318	408	3792	2017	9.3	**4.9**	0.5	1.9	0.48	Proteolysis
CXCL10	3627	115	211	533	1.8	**4.6**	2.5	0.4	4.36	Positive regulation of leukocyte chemotaxis
IL1R2	7850	53	435	241	8.3	**4.6**	0.6	1.8	0.49	Immune response
ICAM1	3383	47	367	213	7.9	**4.6**	0.6	1.7	0.52	T cell antigen processing and presentation
BIRC3	330	27	147	114	5.4	**4.2**	0.8	1.3	0.72	Toll-like receptor signaling pathway
SGPP2	—	51	297	206	5.8	**4.0**	0.7	1.4	0.63	Phospholipid metabolic process

Table 3. Cont.

		Mean of Normalized Expression, Duplicates			Expression Ratios				Scaled eepR		
C15orf48	84419	26	92	99	3.5	**3.8**	1.1	0.9	1.11	Mitochondrion
JMJD4	65094	38	56	146	1.5	**3.8**	2.6	0.4	5.88	Protein binding
C6orf132	647024	42	140	159	3.3	**3.8**	1.1	0.9	1.19	Unknown
S100A7	6278	76	168	288	2.2	**3.8**	1.7	0.6	2.31	Response to reactive oxygen species
KMO	8564	27	107	99	3.9	**3.6**	0.9	1.1	0.89	Metabolic process
EFNA1	1942	278	1973	985	7.1	**3.5**	0.5	2.0	0.42	Negative regulation of transcription from RNA polymerase II promoter
FAM20C	56975	148	528	525	3.6	**3.5**	1.0	1.0	0.99	Phosphorylation
CXCL8	3576	41	2346	145	56.6	**3.5**	0.1	16.2	0.04	Inflammatory cytokine
KMO	8564	28	121	96	4.4	**3.5**	0.8	1.3	0.74	Metabolic process

Nine genes in bold also appear in Table 1: "Greatest expression increase in cells treated with WT vs. mock secretomes".

Table 4. Twenty-five genes with the greatest expression decrease in cells treated with *eepR* vs. mock secretomes.

		Mean of Normalized Expression, Duplicates			Expression Ratios				Scaled eepR	
Gene Symbol	Entrez Gene number	LB control	WT *Serratia*	*eepR* mutant	WT/cont	*eepR*/cont	*eepR*/WT	WT/*eepR*	$\frac{(eppR-cont)}{(WT-cont)}$	Biological Function
CTGF	1490	1245	55	35	0.04	**0.03**	0.64	1.6	−1.02	Cartilage condensation
RP4-813F11.4	—	146	19	13	0.13	**0.09**	0.69	1.5	−1.05	Unknown
AOC3	8639	84	16	8	0.19	**0.10**	0.51	2.0	−1.11	Cell adhesion
SERPINE1	5054	81	42	9	0.52	**0.11**	0.22	4.7	−1.86	Regulation of mRNA stability
TXNIP	10628	2659	64	320	0.02	**0.12**	5.01	0.2	−0.90	Negative regulation of transcription from RNA polymerase II
THAP2	83591	88	15	11	0.17	**0.13**	0.74	1.4	−1.05	Nucleic acid binding
SLC6A13	6540	122	46	18	0.38	**0.15**	0.39	2.6	−1.37	Transport
RFPL3S	10737	31	11	6	0.35	**0.18**	0.53	1.8	−1.25	Unknown
242708_at	—	44	9	8	0.20	**0.19**	0.95	1.1	−1.01	Unknown
SLC26A7	115111	24	4	5	0.14	**0.19**	1.31	0.8	−0.95	Gastric acid secretion
EGR3	1960	567	602	108	1.06	**0.19**	0.18	5.6	−12.90	Positive regulation of endothelial cell proliferation
SERTAD4	56256	40	14	8	0.35	**0.20**	0.58	1.8	−1.23	Unknown
236865_at	—	117	7	24	0.06	**0.21**	3.55	0.3	−0.84	Unknown
ARRDC4	91947	1338	96	285	0.07	**0.21**	2.97	0.3	−0.85	Temperature homeostasis
MYEF2	50804	37	5	8	0.14	**0.23**	1.65	0.6	−0.90	Transcription, DNA-dependent
RYBP	23429	63	31	15	0.49	**0.24**	0.49	2.1	−1.49	Negative regulation of transcription from RNA polymerase II promoter
238548_at	238548_at	44	19	11	0.43	**0.25**	0.59	1.7	−1.31	Unknown
LOC100130705	100130705	67	29	17	0.43	**0.26**	0.60	1.7	−1.30	Unknown
CYR61	3491	5396	1506	1419	0.28	**0.26**	0.94	1.1	−1.02	Regulation of cell growth
ZBTB1	22890	396	217	108	0.55	**0.27**	0.50	2.0	−1.61	Transcription, DNA-dependent

Table 4. Cont.

		Mean of Normalized Expression, Duplicates			Expression Ratios				Scaled eepR	
FOS	2353	425	496	117	1.17	**0.28**	0.24	4.2	−4.34	Toll-like receptor signaling pathway
BC034636 /// CTB-113P19.4	—	53	18	15	0.34	**0.28**	0.81	1.2	−1.10	Unknown
ANGPTL4	51129	393	89	111	0.23	**0.28**	1.25	0.8	−0.93	Angiogenesis
UQCRB	7381	47	14	14	0.30	**0.30**	1.02	1.0	−0.99	Oxidative phosphorylation
C1orf52	148423	171	56	52	0.33	**0.30**	0.93	1.1	−1.04	Unknown

Eight genes in bold also appear in Table 2: "Greatest expression decrease in cells treated with WT vs. mock secretomes".

Table 5. Twenty-five genes with the highest scaled *eepR* values (i.e., relatively little effect of WT, relatively large increase by *eepR*).

Gene Symbol	Entrez Gene number	Mean of Normalized Expression, Duplicates			Expression Ratios				Scaled eepR	Biological Function
		LB control	WT *Serratia*	*eepR* mutant	WT/cont	*eepR*/cont	*eepR*/WT	WT/*eepR*	$\frac{(eppR-cont)}{(WT-cont)}$	
TOMM40L	84134	72	71	237	0.99	3.29	3.33	0.3	235.4	Transport
ARL11	115761	18	19	51	1.01	2.75	2.72	0.4	161.0	Intracellular protein transport
IGFL1	374918	170	173	420	1.02	2.47	2.43	0.4	83.3	Protein binding
227356_at	—	109	112	182	1.02	1.66	1.63	0.6	30.8	Unknown
TMEM177	80775	125	120	221	0.96	1.76	1.84	0.5	18.5	Membrane
TRIM14	9830	221	212	376	0.96	1.70	1.77	0.6	17.8	Protein binding
ZSCAN16	80345	54	52	97	0.95	1.78	1.86	0.5	17.2	Transcription, DNA-dependent
RITA1	84934	80	75	157	0.94	1.97	2.10	0.5	15.1	Intracellular protein transport
KRT34 /// LOC100653049	3885 /// 100653049	202	220	463	1.09	2.29	2.11	0.5	14.8	Epidermis development
CTSC	1075	69	64	135	0.93	1.97	2.11	0.5	13.7	T cell mediated cytotoxicity
FAM13B	51306	128	121	218	0.95	1.70	1.80	0.6	13.1	Signal transduction
CCDC8	83987	70	58	215	0.83	3.08	3.73	0.3	12.1	Negative regulation of phosphatase activity
KIAA1586	57691	34	37	65	1.08	1.91	1.77	0.6	11.9	Nucleic acid binding
COG8 /// PDF	64146 /// 84342	199	217	384	1.09	1.93	1.77	0.6	10.7	Translation
MTRR	4552	418	449	708	1.07	1.69	1.58	0.6	9.4	Sulfur amino acid metabolic process
SLC35F6	54978	125	141	269	1.13	2.15	1.91	0.5	9.1	Establishment of mitotic spindle orientation
CXCL11	6373	121	100	287	0.83	2.38	2.87	0.3	8.1	Positive regulation of leukocyte chemotaxis
HSD17B1	3292	83	103	234	1.24	2.82	2.28	0.4	7.6	Lipid metabolic process
LOC284926	284926	8	13	44	1.63	5.55	3.41	0.3	7.2	Unknown
NOP56	10528	206	233	384	1.13	1.87	1.65	0.6	6.5	rRNA processing
* FAM86B1	*55199	32	26	65	0.82	2.07	2.51	0.4	6.1	Unknown
JMJD4	65094	38	56	146	1.48	3.82	2.58	0.4	5.9	Protein binding
PPAPDC2	403313	67	85	156	1.26	2.33	1.84	0.5	5.0	Metabolic process
AIMP2	7965	1059	966	1505	0.91	1.42	1.56	0.6	4.8	Translation
ZNF165	7718	184	220	358	1.20	1.95	1.63	0.6	4.8	Transcription, DNA-dependent

* full annotation: FAM86B1 /// FAM86B2 /// FAM86C1 /// FAM86DP /// FAM86FP /// FAM86KP: 55199 /// 85002 /// 653113 /// 653333 /// 692099 /// 100287013.

Table 6. Twenty-five genes with the lowest scaled *eepR* values (i.e., relatively little effect of WT, relatively large decrease by *eepR*).

Gene Symbol	Entrez Gene number	Mean of Normalized Expression, Duplicates			Expression Ratios				Scaled *eepR*	Biological Function
		LB control	WT *Serratia*	*eepR* mutant	WT/cont	*eepR*/cont	*eepR*/WT	WT/*eepR*	(*eepR*—cont) / |(WT—cont)|	
NUFIP2	57532	1790	1786	1204	1.00	0.67	0.67	1.5	−144.7	Protein binding
ZFP36L2	678	3551	3536	2305	1.00	0.65	0.65	1.5	−83.9	Regulation of transcription, DNA dependent
TUFT1	7286	1322	1329	818	1.01	0.62	0.62	1.6	−74.7	Protein binding
PARD6B	84612	495	491	230	0.99	0.46	0.47	2.1	−59.7	Protein complex assembly
GPR157	80045	364	359	208	0.99	0.57	0.58	1.7	−33.0	Signal transduction
ARPC5L	81873	1049	1062	602	1.01	0.57	0.57	1.8	−32.4	Regulation of actin filament polymerization
JUN	3725	1312	1338	621	1.02	0.47	0.46	2.2	−27.1	Angiogenesis
DUSP6	1848	3965	3858	1638	0.97	0.41	0.42	2.4	−21.6	Inactivation of MAPK activity
CD274	29126	373	386	184	1.04	0.49	0.48	2.1	−14.4	Immune response
EGR3	1960	567	602	108	1.06	0.19	0.18	5.6	−12.9	Positive regulation of endothelial cell proliferation
1555897_at	—	89	85	47	0.96	0.53	0.55	1.8	−11.9	Unknown
CHMP1B	57132	220	227	145	1.03	0.66	0.64	1.6	−11.4	Cytokinesis
FHL2	2274	2117	2058	1460	0.97	0.69	0.71	1.4	−11.3	Negative regulation of transcription from RNA polymerase II promoter
E2F7	144455	1501	1411	612	0.94	0.41	0.43	2.3	−10.0	Negative regulation of transcription from RNA polymerase II promoter
SLC2A14 /// SLC2A3	6515 /// 144195	358	342	203	0.96	0.57	0.59	1.7	−9.8	Carbohydrate metabolic process
PHF13	148479	631	661	347	1.05	0.55	0.52	1.9	−9.5	Mitotic cell cycle
JAG1	182	3839	3973	2726	1.03	0.71	0.69	1.5	−8.3	Angiogenesis
SERTAD1	29950	1254	1334	653	1.06	0.52	0.49	2.0	−7.5	Regulation of cyclin-dependent protein serine/threonine kinase activity
KIAA0907	22889	1868	1755	1034	0.94	0.55	0.59	1.7	−7.4	Unknown
SOS1	6654	442	467	268	1.06	0.61	0.58	1.7	−7.1	Apoptotic process
C16orf72	29035	2072	2169	1411	1.05	0.68	0.65	1.5	−6.8	Unknown
RND3	390	1928	1794	1126	0.93	0.58	0.63	1.6	−6.0	GTP catabolic process
SMAD7	4092	271	290	158	1.07	0.58	0.54	1.8	−5.7	Negative regulation of transcription from RNA polymerase II promoter
ADAMTS6	11174	171	187	86	1.10	0.50	0.46	2.2	−5.1	Proteolysis
FZD7	8324	63	60	31	0.89	0.45	0.50	2.0	−5.0	Wnt signaling

Table 7. Significantly ($p < 0.05$) enriched canonical pathways which respond to WT *S. marcescens* stimulus.

Canonical Pathway.	−log(p-Value)	Number Genes up-Regulated	Number Genes down-Regulated	Total Genes in Pathway
IL-6 Signaling	8.4	20	1	116
Toll-like Receptor Signaling	7.9	16	1	72
NF-kB Signaling	7.0	22	1	164
Colorectal Cancer Metastasis Signaling	5.4	20	1	230
PPAR Signaling	5.0	14	1	90
TREM1 Signaling	4.9	12	1	69
HMGB1 Signaling	4.8	15	1	118
Acute Phase Response Signaling	4.6	18	1	166
Role of Pattern Recognition Receptors in Recognition of Bacteria and Viruses	4.2	13	1	118
Cholecystokinin/Gastrin-mediated Signaling	3.3	11	1	99
B Cell Activating Factor Signaling	3.1	7	1	40
LXR/RXR Activation	3.1	12	1	120
Pancreatic Adenocarcinoma Signaling	3.0	10	1	106
Glioma Invasiveness Signaling	2.8	7	1	57
Cell Cycle: G2/M DNA Damage Checkpoint Regulation	2.6	1	1	49
NF-kB Activation by Viruses	2.1	7	1	73
NRF2-mediated Oxidative Stress Response	2.0	12	1	175
CXCL8 Signaling	1.9	13	1	
Tec Kinase Signaling	1.8	10	2	183
MIF Regulation of Innate Immunity	1.8	5	0	150
iNOS Signaling	1.6	5	0	39
Antioxidant Action of Vitamin C	1.6	8	0	43
PPARα/RXRα Activation	1.5	12	0	91
Phospholipase C Signaling	1.3	13	1	165

The two pathways in the bold text were also significantly stimulated by *eepR* secretomes with the same −log(p values) found for WT secretomes. Note: This indicates that most immune pathways in this table modulated by WT secretome treatment are not modulated by *eepR* secretome treatment.

Interestingly, when we examined our results in the context of scaled *eepR* (*eepR*—control/WT—control), there were also several genes where the expression difference was greater than 10-fold in *eepR*-treated cells in comparison to WT. In particular, there were differences in genes involved in intracellular protein transport, protein binding, transcription, nucleic acid binding, and translation (Table 5). Genes with the lowest expression in the scaled *eepR* response were involved in protein complex assembly, protein binding, signal transduction, actin filament polymerization, inactivation of MAPK activity, and negative regulation of transcription (Table 6).

From our microarray results, we chose genes to validate by qRT-PCR that are known mediators of response to infection and corneal wound healing, involved in cellular signaling, motility, actin binding, and cellular division/membrane organization, and had at least a 2-fold difference when comparing WT to *eepR*-treated HCLEs [17,20,21,27]. Overall, our qRT-PCR results validated changes observed with the microarray, including that the *eepR*-treated HCLEs in most cases had a lower fold change in proinflammatory gene expression (Figure 1, Table 1). We note that, when assayed by qRT-PCR, nine out of the twelve genes show a greater response to WT treatment than they do by microarray analysis, consistent with the greater sensitivity and wider dynamic range of qRT-PCR.

2.2. Bacterial Secretomes Influence Corneal Epithelial Cell Lipid Metabolism

In addition to evidence of EepR playing a role in producing inflammatory markers, microarray analysis revealed alterations in pathways associated with lipid metabolism and signaling. These pathways include ceramide biosynthesis, ceramide signaling, and Sphingosine-1-phosphate receptor signaling with a valid increases in CERS2 (1.5-fold), S1PR3 (2.3-fold), SPHK1 (1.8-fold), and SPTLC2 (2.5-fold) genes by cells treated with wild-type, but not *eepR* secretomes. Increased CERS2 expression observed in the microarray was confirmed by qRT-PCR (Figure 1).

To further verify the alteration in producing compounds associated with the lipid pathways implicated in the microarray data and to gain insight into the corneal epithelial cell response to enteric bacteria, small molecule metabolomic analysis was performed on HCLE cells exposed to secretomes derived from WT and the *eepR* mutant. Consistent with the results of the microarray analysis and qRT-PCR, the metabolomic analysis identified changes in markers involved in lipid metabolism (Figure 2, Supplementary Tables S1 and S2). There were significant increases in metabolites for lipid metabolism for *S. marcescens* WT-treated HCLEs, including sphingosine, phosphoethanolamine (Figure 2), as well as linoleate, eicosapentaenoate, docosapentaenoate, docosahexaenoate, and myristate (Table S1). Together, these data indicate that *S. marcescens* secreted factors have a major impact on human corneal cells, including increased expression of inflammatory and lipid metabolism pathways, and that *S. marcescens* requires EepR for these effects.

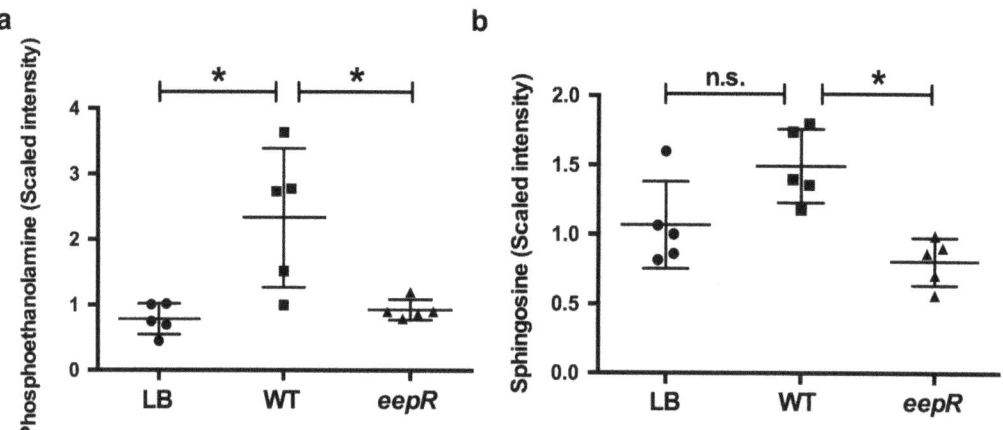

Figure 2. Metabolomic analysis demonstrates alteration of sphingosine and lipid metabolism in corneal cells challenged by *S. marcescens* secretomes. HCLE cells were treated with LB, WT, or *eepR* secretomes for 24 h. Mean and SD (n = 5) of relative amounts of (**a**) phosphoethanolamine and (**b**) sphingosine. Circles = LB (mock treatment), squares = WT, and triangles = *eepR* mutant treated HCLE cells. Asterisks (*) indicate $p < 0.05$ by one-way ANOVA with Tukey's post-hoc analysis. n.s., not significant.

3. Discussion

S. marcescens EepR, a master transcriptional regulator of secreted enzymes and secondary metabolites, plays an important role in hemolysis, pigment production, swarming motility, and contributes to bacterial proliferation in the cornea. A previous study demonstrated the importance of the *S. marcescens* transcription factor EepR in the regulation of protease production, corneal cell-induced cytotoxicity, and its ability to induce the proinflammatory cytokine IL-1β [17]. Because of its involvement in ocular host-pathogen response, we sought to determine differences in gene expression profiles in *eepR*-treated corneal cells in comparison to WT. Interestingly, genes with the greatest expression in *eepR* mutant-treated corneal cells compared to WT-treated cells were those involved in intracellular transport, protein binding, cellular component movement, cell adhesion, and membrane-related functions (Table 5), suggesting deletion of EepR promotes cell migration and wound healing. Consistently, *eepR*-treated cells were found to regulate lipid metabolic process, transcription, and intracellular protein transport (Table 5) and activate the MAPK pathway, which has been demonstrated to promote cell migration [28]. In contrast, WT-treated cells were found to inactivate the MAPK pathway (Table 6), which is consistent with its wound inhibitory phenotype [16].

The effect of bacteria on human corneal epithelial cells is of interest because bacteria cause the majority of corneal ulcers [29]. A limited number of studies have examined the impact of *P. aeruginosa* and *S. aureus* on the corneal transcriptomic response [20–23], but these have not been done with bacteria of the Enterobacterales order. Bacteria, such as *Klebsiella*, *Proteus*, and *Serratia*, cause a significant number of ocular infections [30]. There is a unique immunological response of the cornea, being an immune-privileged site. Chidambaram et al. compared gene expression profiles of corneal tissues from microbial keratitis patients infected with *Streptococcus pneumoniae*, *P. aeruginosa*, *Fusarium sp.*, and *Aspergillus sp.* to normal corneal tissue from cadavers [20]. In agreement with our own data, they found increased expression of the proinflammatory markers MMP9, MMP1, IL-1β, and TNF with the greatest expression observed in MMP9. In addition to the previously mentioned markers, they also found increases in MMP7, MMP10, MMP12, TLR2, and TLR4, all markers known to promote inflammation and immune recognition [20]. Our data also found a 2.2-fold increase in expression of TLR2 in WT *versus eepR* mutant-treated HCLEs, but no significant changes in TLR4 expression. Microarray gene expression levels for TLR4 were low, but detectable for all conditions in our study. However, expression of TLR4 in corneal epithelial cells has been previously demonstrated to be reduced [31,32], and could explain why our results differed from Chidambaram et al.

The *S. marcescens*-induced proinflammatory gene response reported here was consistent with a study by Hume et al. [27], who used ELISA to explore the cytokine response of human corneal cells and polymorphonuclear monocytes (PMNs) to clinical isolates of *S. marcescens*. Though they found strain differences in cytokine response, there was an overall positive trend in activation of TNFα, IL-6, and CXCL8 after 4 h of exposure to bacteria which was similar to our results after 5 h of exposure [27].

The impact of living *Pseudomonas aeruginosa* upon the transcriptome of murine corneas has been explored by Gao et al. [21]. They reported upregulation of Krt16, MMP10, MMP13, S100A8, Stfna111, and S100A9 genes with an even greater increase in the genes involved in antimicrobial peptide production S100A8 and S100A9, when mice were pretreated with flagellin [21]. Our results were not as striking for S100A8 and S100A9, but did demonstrate a 2-fold increase in WT-treated HCLES in comparison to *eepR*. Huang et al. used murine corneas infected with *P. aeruginosa* and demonstrated upregulation of proinflammatory markers GM-CSF, ICAM1, IL1α, IL-1β, IL-6, TNFα, MMP9, MMP10, and MMP13 in accordance with our results [22]. In addition to the previously mentioned genes, we also observed upregulation of proinflammatory markers CCL20, CERS2, CXCL1, CXCL8, and MMP1.

An elegant study by Heimer et al. used a well-defined reference strain of the gram positive bacteria *S. aureus* to examine corneal epithelial cell responses to bacteria [23]. They evaluated the effect of an isogenic *agr sarA* double mutant of *S. aureus* that has similar defects as our *eepR* mutant in reduced secretion of virulence factors [23]. After treating human corneal cells with *S. aureus*, highly increased expression of proinflammatory markers CCL20, CSF2, CXCL1, IL-6, CXCL8, and TNFα was observed. These results are in agreement with our own, with the only major notable difference being that the gene most induced by *S. marcescens* WT bacteria was CXCL8, a neutrophil chemoattractant important for neutrophil migration to the site of infection and clearance of bacteria, whereas *S. aureus* most induced CCL20 a chemokine with antibacterial properties [33]—the third most highly induced gene in our study. In sharp contrast to their study, while the *S. aureus agr sarA* double mutant caused relatively little change in host response compared to the

WT *S. aureus*, the *S. marcescens eepR* mutant was strikingly less able than the WT to induce expression of proinflammatory genes. Another notable difference is that some of the signal transduction factors upregulated by *S. aureus* were not affected by *S. marcescens*, notably the plasminogen activator inhibitor SERPINB2 that is involved in macrophage function and cell migration [34], and the glycoprotein STC1 that is involved in angiogenesis and wound healing [35].

Matrix metalloproteinases (MMPs) are enzymes that function in immune responses to infection in addition to numerous other roles. MMPs are involved in recruiting white blood cells, chemokine and cytokine responses, and cell matrix remodeling [36]. In our study, numerous matrix metalloproteases were upregulated >2-fold by *S. marcescens*, including MMP1, 9, 10, 13, 14, 16, 19, 28, but a similar trend was not described in *S. aureus* challenged cells [23].The different pathogen associated molecular patterns produced by the bacteria and the challenge with whole *S. aureus* versus *S. marcescens* secretomes (which include flagella and LPS) may account for some of the differences observed. Nevertheless, the *S. marcescens* EepR protein had a much larger role than the *S. aureus* SarA transcription factor and Agr quorum sensing system in affecting the corneal epithelial cell transcriptional response.

The reason for which *eepR* mutants confer such a different transcriptional response compared to the WT is not clear at this time. The *eepR* mutant is defective in the secretion of metalloproteases, such as serralysin and SlpB [17]. Serralysin, also called the 56-kDa protease, was shown in experimental models to have an impact on the immune system, rendering mouse lungs much more susceptible to influenza infection [37]. The protease was shown to increase vascular permeability by activation of the Hageman factor-kallikrein-kinin system [38]. Further studies will evaluate the role of EepR regulated bacterial metalloproteases in corneal wound healing.

Our microarray and qRT-PCR data suggested differences for expression of genes involved in the lipid metabolism pathway for corneal cells exposed to WT, but not *eepR* mutant secretomes. This data was validated using metabolomics approaches and indicated that the changes in transcription yielded measurable differences in the molecules involved in the altered pathways. Bioactive sphingolipids, such as those with altered expression shown here, like ceramide and sphingosine 1-phosphate, are known signaling molecules that mediate wound healing in many tissues [39], and likely play a different role in corneal responses. These data indicate the importance of a single bacterial transcription factor in dictating the corneal cell response as measured through transcriptomic and metabolomic analysis. The findings and their implications should be discussed in the broadest context possible.

4. Materials and Methods

4.1. Bacterial Growth Conditions and Media

S. marcescens cultures were grown in lysogeny broth (LB) [40] at 30 °C with shaking. Bacteria free secretomes of *S. marcescens* WT and *eepR* were prepared by normalizing overnight cultures to OD_{600} = 2.0 and removing the bacteria by centrifugation at 14,000 rpm for two minutes followed by filtration through a 0.22 µm filter.

4.2. Microarray

HCLE cell line was obtained from Ilene Gipson [41], and were maintained in KSFM media as previously described [16]. Cells were seeded into 12 well plates at a density of 1.5×10^5 cells per well. Secretomes were prepared as described above and added to HCLE cells at the same dosage (500 µL into 1 mL KSFM) and incubated for 5 h at 37 °C + 5% CO_2. HCLE cells were washed 3 times with phosphate buffered saline (PBS) and stored in 5 volumes of RNA*later* (Sigma-Aldrich, St. Louis, MO, USA) at 4 °C until used. RNA was extracted with a GenElute Mammalian total RNA miniprep kit (Sigma-Aldrich), treated with 1 unit of RQ1 Dnase (Promega, Madison, WI, USA) for 30 min at 37 °C, and quantified by Nanodrop (Thermo Scientific | Thermo Fisher Scientific, Waltham, MA, USA). 500 ng samples of total RNA were processed using an Affymetrix 3′-IVT Express kit (Affymetrix,

Santa Clara, CA, USA) and yielded 43.2 ± 14.4 μg of biotinylated cRNA (mean ± SD, n = 5), with one outlier of 7 μg. Twenty μg of biotinylated cRNA was hybridized to Affymetrix U133 Plus 2.0 GeneChips (catalog #900470). The GeneChips were developed and scanned using an Affymetrix GeneChip 3000 Array Scanner.

The resultant DAT files were consolidated to CEL files, which were analyzed with Affymetrix GCOS v1.4 software, using default parameters. Numerical data and the software flags for Presence/Absence and for significant pairwise changes were transferred to Microsoft Excel. Of the 54,675 panels (unique sequence targets) on the microarray, 26,162 showed no detectable expression in any sample and omitted further consideration. Of the remaining 28,513 panels, the 22,553 (79%) which showed consistent detectable expression in the duplicate samples of at least one experimental group were taken for analysis. Of these, 621 panels (2.8%) showed a significant 2-fold difference between duplicates and were rejected as unreliable. For the reliable 21,932 panels, the ratio (mean (WT − treated)/mean (untreated)) was calculated. This ratio represented a valid change if:Both samples in the higher-expressing group reported Present (i.e., detectable target sequence), all four pairwise comparisons between groups showed significant changes using the GCOS software, and the groups did not overlap.

4.3. Quantitative Reverse Transcriptase PCR (qPCR)

RNA was extracted as described above and concentrated using an RNA Clean and Concentration kit (Zymo Research, Irvine, CA, USA). All samples were normalized with nuclease free water to a concentration of 50 ng/μL. 250 μg of RNA was synthesized into cDNA using Superscript III reverse transcriptase (Invitrogen | Thermo Fisher Scientific, Waltham, MA, USA) as previously described [19]. To identify any genomic DNA contamination, non-template controls of each RNA sample were also prepared and verified by reverse transcriptase PCR (RT-PCR) using GAPDH primers [42]. All contaminated samples were discarded. Quantitative reverse transcriptase PCR (qRT-PCR) was performed using Sybr green reagent (Applied Biosystems | Thermo Fisher Scientific, Waltham, MA, USA) using primers for CCL20, CERS2, CSF2, ICAM-1, IL-1α, IL-1β, IL-6, IL-8, MMP1, MMP9, TNFα [42–52]. All gene reactions were normalized to GAPDH [42], and analyzed using the $\Delta\Delta CT$ method. All experiments were performedat least three independent times.

4.4. Metabolomics

One sample containing 100 μL of LB (mock) and five 100 μL samples each of WT and *eepR* mutant *were* collected and stored at −80 °C. All samples were collected in two independent harvests on two different days and shipped on dry ice to Metabolon Inc. for small molecule analysis. Samples were prepared using an automated MicroLab STAR® system (The Hamilton Company, Allston, MA, USA) using a proprietary series of organic and aqueous extractions. The prepared extract was then divided into two fractions, one for analysis by liquid chromatography and one for analysis by gas chromatography. Samples were then placed in a TurboVap® (Biotage, Uppsala, Sweden) to remove the organic solvent. Each sample was frozen and dried under vacuum and prepared for liquid chromatography mass spectrometry (LC/MS) or gas chromatography mass spectrometry analysis. Library entries of purified standards or recurrent unknown entities were used to identify compounds. Matches for each sample were verified and corrected as needed.

4.5. Statistical Analysis

Student's *t*-test and one-way ANOVA with post hoc statistical tests were performed using GraphPad Prism statistical software version 6.0. For metabolomics analysis, Welch's *t*-tests using pairwise comparisons were performed for statistical analysis. Significance for all statistical tests was determined at $p < 0.05$.

Supplementary Materials: The following are available online at https://www.mdpi.com/article/10.3390/antibiotics10070770/s1, Table S1: Metabolomics analysis of WT and *eepR* secretome-treated HCLE, Table S2: Metabolomics data.

Author Contributions: Conceptualization, K.M.B. and R.M.Q.S.; methodology, K.M.B., S.A.K.H. and R.M.Q.S.; software, S.A.K.H.; investigation, K.M.B. and S.A.K.H.; data curation, K.M.B., S.A.K.H. and R.M.Q.S.; writing—original draft preparation, K.M.B.; writing—review and editing, K.M.B., S.A.K.H. and R.M.Q.S.; funding acquisition, R.M.Q.S. All authors have read and agreed to the published version of the manuscript.

Funding: This work was supported by Research to Prevent Blindness (unrestricted funds), the Eye and Ear Foundation of Pittsburgh, National Institute of Health grants P30EY08098 (to Department of Ophthalmology), F32EY024785 (to K.M.B.), T32EY017271 (to K.M.B.), and R01EY027331 (to R.M.Q.S.).

Data Availability Statement: Microarray data was deposited to NCBI gene expression Omnibus (GEO accession number GSM1832614). Metabolomic data is supplied in Table S2.

Conflicts of Interest: The authors declare no conflict of interest.

References

1. Fleiszig, S.M.; McNamara, N.A.; Evans, D.J. The tear film and defense against infection. *Adv. Exp. Med. Biol.* **2002**, *506*, 523–530.
2. McDermott, A.M. Antimicrobial compounds in tears. *Exp. Eye Res.* **2013**, *117*, 53–61. [CrossRef]
3. Evans, D.J.; Fleiszig, S.M. Why does the healthy cornea resist *Pseudomonas aeruginosa* infection? *Am. J. Ophthalmol.* **2013**, *155*, 961–970. [CrossRef] [PubMed]
4. Metruccio, M.M.E.; Tam, C.; Evans, D.J.; Xie, A.L.; Stern, M.E.; Fleiszig, S.M.J. Contributions of MyD88-dependent receptors and CD11c-positive cells to corneal epithelial barrier function against *Pseudomonas aeruginosa*. *Sci. Rep.* **2017**, *7*, 13829. [CrossRef] [PubMed]
5. Callegan, M.C.; O'Callaghan, R.J.; Hill, J.M. Pharmacokinetic considerations in the treatment of bacterial keratitis. *Clin. Pharmacokinet.* **1994**, *27*, 129–149. [CrossRef] [PubMed]
6. Hazlett, L.D. Role of innate and adaptive immunity in the pathogenesis of keratitis. *Ocul. Immunol. Inflamm.* **2005**, *13*, 133–138. [CrossRef] [PubMed]
7. Pearlman, E.; Sun, Y.; Roy, S.; Karmakar, M.; Hise, A.G.; Szczotka-Flynn, L.; Ghannoum, M.; Chinnery, H.R.; McMenamin, P.G.; Rietsch, A. Host defense at the ocular surface. *Int. Rev. Immunol.* **2013**, *32*, 4–18. [CrossRef] [PubMed]
8. Ruan, X.; Chodosh, J.; Callegan, M.C.; Booth, M.C.; Lee, T.D.; Kumar, P.; Gilmore, M.S.; Pereira, H.A. Corneal expression of the inflammatory mediator CAP37. *Investig. Ophthalmol. Vis. Sci.* **2002**, *43*, 1414–1421.
9. Willcox, M.D. *Pseudomonas aeruginosa* infection and inflammation during contact lens wear: A review. *Optom. Vis. Sci.* **2007**, *84*, 273–278. [CrossRef]
10. Mah-Sadorra, J.H.; Najjar, D.M.; Rapuano, C.J.; Laibson, P.R.; Cohen, E.J. *Serratia* corneal ulcers: A retrospective clinical study. *Cornea* **2005**, *24*, 793–800. [CrossRef]
11. Voelz, A.; Muller, A.; Gillen, J.; Le, C.; Dresbach, T.; Engelhart, S.; Exner, M.; Bates, C.J.; Simon, A. Outbreaks of *Serratia marcescens* in neonatal and pediatric intensive care units: Clinical aspects, risk factors and management. *Int. J. Hyg. Environ. Health* **2010**, *213*, 79–87. [CrossRef]
12. Mahlen, S.D. *Serratia* infections: From military experiments to current practice. *Clin. Microbiol. Rev.* **2011**, *24*, 755–791. [CrossRef]
13. Rahim, K.; Saleha, S.; Zhu, X.; Huo, L.; Basit, A.; Franco, O.L. Bacterial Contribution in Chronicity of Wounds. *Microb. Ecol.* **2017**, *73*, 710–721. [CrossRef]
14. Scales, B.S.; Huffnagle, G.B. The microbiome in wound repair and tissue fibrosis. *J. Pathol.* **2013**, *229*, 323–331. [CrossRef]
15. Scali, C.; Kunimoto, B. An update on chronic wounds and the role of biofilms. *J. Cutan. Med. Surg.* **2013**, *17*, 371–376. [CrossRef]
16. Brothers, K.M.; Stella, N.A.; Hunt, K.M.; Romanowski, E.G.; Liu, X.; Klarlund, J.K.; Shanks, R.M. Putting on the brakes: Bacterial impediment of wound healing. *Sci. Rep.* **2015**, *5*, 14003. [CrossRef]
17. Brothers, K.M.; Stella, N.A.; Romanowski, E.G.; Kowalski, R.P.; Shanks, R.M. EepR Mediates Secreted-Protein Production, Desiccation Survival, and Proliferation in a Corneal Infection Model. *Infect. Immun.* **2015**, *83*, 4373–4382. [CrossRef]
18. Shanks, R.M.Q.; Stella, N.A.; Lahr, R.M.; Aston, M.A.; Brothers, K.M.; Callaghan, J.D.; Sigindere, C.; Liu, X. Suppressor analysis of eepR mutant defects reveals coordinate regulation of secondary metabolites and serralysin biosynthesis by EepR and HexS. *Microbiology* **2017**, *163*, 280–288. [CrossRef]
19. Stella, N.A.; Lahr, R.M.; Brothers, K.M.; Kalivoda, E.J.; Hunt, K.M.; Kwak, D.H.; Liu, X.; Shanks, R.M. *Serratia marcescens* cyclic AMP-receptor protein controls transcription of EepR, a novel regulator of antimicrobial secondary metabolites. *J. Bacteriol.* **2015**, *197*, 2468–2478. [CrossRef]
20. Chidambaram, J.D.; Kannambath, S.; Srikanthi, P.; Shah, M.; Lalitha, P.; Elakkiya, S.; Bauer, J.; Prajna, N.V.; Holland, M.J.; Burton, M.J. Persistence of Innate Immune Pathways in Late Stage Human Bacterial and Fungal Keratitis: Results from a Comparative Transcriptome Analysis. *Front. Cell Infect. Microbiol.* **2017**, *7*, 193. [CrossRef]
21. Gao, N.; Sang Yoon, G.; Liu, X.; Mi, X.; Chen, W.; Standiford, T.J.; Yu, F.S. Genome-wide transcriptional analysis of differentially expressed genes in flagellin-pretreated mouse corneal epithelial cells in response to *Pseudomonas aeruginosa*: Involvement of S100A8/A9. *Mucosal Immunol.* **2013**, *6*, 993–1005. [CrossRef]

22. Huang, X.; Hazlett, L.D. Analysis of *Pseudomonas aeruginosa* corneal infection using an oligonucleotide microarray. *Investig. Ophthalmol. Vis. Sci.* **2003**, *44*, 3409–3416. [CrossRef]
23. Heimer, S.R.; Yamada, A.; Russell, H.; Gilmore, M. Response of corneal epithelial cells to *Staphylococcus aureus*. *Virulence* **2010**, *1*, 223–235. [CrossRef]
24. Shanks, R.M.; Stella, N.A.; Hunt, K.M.; Brothers, K.M.; Zhang, L.; Thibodeau, P.H. Identification of SlpB, a Cytotoxic Protease from *Serratia marcescens*. *Infect. Immun.* **2015**, *83*, 2907–2916. [CrossRef] [PubMed]
25. Brothers, K.M.; Kowalski, R.P.; Tian, S.; Kinchington, P.R.; Shanks, R.M.Q. Bacteria induce autophagy in a human ocular surface cell line. *Exp. Eye Res.* **2018**, *168*, 12–18. [CrossRef]
26. Brothers, K.M.; Stella, N.A.; Shanks, R.M.Q. Biologically active pigment and ShlA cytolysin of Serratia marcescens induce autophagy in a human ocular surface cell line. *BMC Ophthalmol.* **2020**, *20*, 120. [CrossRef]
27. Hume, E.; Sack, R.; Stapleton, F.; Willcox, M. Induction of cytokines from polymorphonuclear leukocytes and epithelial cells by ocular isolates of *Serratia marcescens*. *Ocul. Immunol. Inflamm.* **2004**, *12*, 287–295. [CrossRef]
28. Saika, S.; Okada, Y.; Miyamoto, T.; Yamanaka, O.; Ohnishi, Y.; Ooshima, A.; Liu, C.Y.; Weng, D.; Kao, W.W. Role of p38 MAP kinase in regulation of cell migration and proliferation in healing corneal epithelium. *Investig. Ophthalmol. Vis. Sci.* **2004**, *45*, 100–109. [CrossRef] [PubMed]
29. Al-Mujaini, A.; Al-Kharusi, N.; Thakral, A.; Wali, U.K. Bacterial keratitis: Perspective on epidemiology, clinico-pathogenesis, diagnosis and treatment. *Sultan Qaboos Univ. Med. J.* **2009**, *9*, 184–195.
30. Teweldemedhin, M.; Gebreyesus, H.; Atsbaha, A.H.; Asgedom, S.W.; Saravanan, M. Bacterial profile of ocular infections: A systematic review. *BMC Ophthalmol.* **2017**, *17*, 212. [CrossRef]
31. Zhang, J.; Kumar, A.; Wheater, M.; Yu, F.S. Lack of MD-2 expression in human corneal epithelial cells is an underlying mechanism of lipopolysaccharide (LPS) unresponsiveness. *Immunol. Cell Biol.* **2009**, *87*, 141–148. [CrossRef] [PubMed]
32. Ueta, M.; Nochi, T.; Jang, M.H.; Park, E.J.; Igarashi, O.; Hino, A.; Kawasaki, S.; Shikina, T.; Hiroi, T.; Kinoshita, S.; et al. Intracellularly expressed TLR2s and TLR4s contribution to an immunosilent environment at the ocular mucosal epithelium. *J. Immunol.* **2004**, *173*, 3337–3347. [CrossRef] [PubMed]
33. Schutyser, E.; Struyf, S.; Van Damme, J. The CC chemokine CCL20 and its receptor CCR6. *Cytokine Growth Factor Rev.* **2003**, *14*, 409–426. [CrossRef]
34. Shea-Donohue, T.; Zhao, A.; Antalis, T.M. SerpinB2 mediated regulation of macrophage function during enteric infection. *Gut Microbes* **2014**, *5*, 254–258. [CrossRef] [PubMed]
35. Guo, F.; Li, Y.; Wang, J.; Li, Y.; Li, Y.; Li, G. Stanniocalcin1 (STC1) Inhibits Cell Proliferation and Invasion of Cervical Cancer Cells. *PLoS ONE* **2013**, *8*, e53989. [CrossRef]
36. Elkington, P.T.; O'Kane, C.M.; Friedland, J.S. The paradox of matrix metalloproteinases in infectious disease. *Clin. Exp. Immunol.* **2005**, *142*, 12–20. [CrossRef]
37. Akaike, T.; Molla, A.; Ando, M.; Araki, S.; Maeda, H. Molecular mechanism of complex infection by bacteria and virus analyzed by a model using serratial protease and influenza virus in mice. *J. Virol.* **1989**, *63*, 2252–2259. [CrossRef]
38. Kamata, R.; Yamamoto, T.; Matsumoto, K.; Maeda, H. A serratial protease causes vascular permeability reaction by activation of the Hageman factor-dependent pathway in guinea pigs. *Infect. Immun.* **1985**, *48*, 747–753. [CrossRef]
39. Shea, B.S.; Tager, A.M. Sphingolipid regulation of tissue fibrosis. *Open Rheumatol. J.* **2012**, *6*, 123–129. [CrossRef]
40. Bertani, G. Studies on lysogenesis. I. The mode of phage liberation by lysogenic *Escherichia coli*. *J. Bacteriol.* **1951**, *62*, 293–300. [CrossRef]
41. Gipson, I.K.; Spurr-Michaud, S.; Argueso, P.; Tisdale, A.; Ng, T.F.; Russo, C.L. Mucin gene expression in immortalized human corneal-limbal and conjunctival epithelial cell lines. *Investig. Ophthalmol. Vis. Sci.* **2003**, *44*, 2496–2506. [CrossRef]
42. Dos Santos, A.; Balayan, A.; Funderburgh, M.L.; Ngo, J.; Funderburgh, J.L.; Deng, S.X. Differentiation Capacity of Human Mesenchymal Stem Cells into Keratocyte Lineage. *Investig. Ophthalmol. Vis. Sci.* **2019**, *60*, 3013–3023. [CrossRef]
43. Nozato, K.; Fujita, J.; Kawaguchi, M.; Ohara, G.; Morishima, Y.; Ishii, Y.; Huang, S.K.; Kokubu, F.; Satoh, H.; Hizawa, N. IL-17F Induces CCL20 in Bronchial Epithelial Cells. *J. Allergy* **2011**, *2011*, 587204. [CrossRef]
44. Erez-Roman, R.; Pienik, R.; Futerman, A.H. Increased ceramide synthase 2 and 6 mRNA levels in breast cancer tissues and correlation with sphingosine kinase expression. *Biochem. Biophys. Res. Commun.* **2010**, *391*, 219–223. [CrossRef]
45. Sun, Y.; Guo, Q.M.; Liu, D.L.; Zhang, M.Z.; Shu, R. In vivo expression of Toll-like receptor 2, Toll-like receptor 4, CSF2 and LY64 in Chinese chronic periodontitis patients. *Oral Dis.* **2010**, *16*, 343–350. [CrossRef]
46. Schultz, K.R.; Klarnet, J.P.; Gieni, R.S.; HayGlass, K.T.; Greenberg, P.D. The role of B cells for in vivo T cell responses to a Friend virus-induced leukemia. *Science* **1990**, *249*, 921–923. [CrossRef]
47. Glushakova, O.; Kosugi, T.; Roncal, C.; Mu, W.; Heinig, M.; Cirillo, P.; Sanchez-Lozada, L.G.; Johnson, R.J.; Nakagawa, T. Fructose induces the inflammatory molecule ICAM-1 in endothelial cells. *J. Am. Soc. Nephrol.* **2008**, *19*, 1712–1720. [CrossRef]
48. Dabkeviciene, D.; Sasnauskiene, A.; Leman, E.; Kvietkauskaite, R.; Daugelaviciene, N.; Stankevicius, V.; Jurgelevicius, V.; Juodka, B.; Kirveliene, V. mTHPC-mediated photodynamic treatment up-regulates the cytokines VEGF and IL-1alpha. *Photochem. Photobiol.* **2012**, *88*, 432–439. [CrossRef]
49. Kulik, T.J. Inhaled nitric oxide in the management of congenital heart disease. *Curr. Opin. Cardiol.* **1996**, *11*, 75–80. [CrossRef]
50. Morris, M.C.; Gilliam, E.A.; Button, J.; Li, L. Dynamic modulation of innate immune response by varying dosages of lipopolysaccharide (LPS) in human monocytic cells. *J. Biol. Chem.* **2014**, *289*, 21584–21590. [CrossRef]

51. Huntington, J.T.; Shields, J.M.; Der, C.J.; Wyatt, C.A.; Benbow, U.; Slingluff, C.L., Jr.; Brinckerhoff, C.E. Overexpression of collagenase 1 (MMP-1) is mediated by the ERK pathway in invasive melanoma cells: Role of BRAF mutation and fibroblast growth factor signaling. *J. Biol. Chem.* **2004**, *279*, 33168–33176. [CrossRef] [PubMed]
52. Safranek, J.; Pesta, M.; Holubec, L.; Kulda, V.; Dreslerova, J.; Vrzalova, J.; Topolcan, O.; Pesek, M.; Finek, J.; Treska, V. Expression of MMP-7, MMP-9, TIMP-1 and TIMP-2 mRNA in lung tissue of patients with non-small cell lung cancer (NSCLC) and benign pulmonary disease. *Anticancer Res.* **2009**, *29*, 2513–2517. [PubMed]

Article

Clearance of Gram-Negative Bacterial Pathogens from the Ocular Surface by Predatory Bacteria

Eric G. Romanowski [1], Shilpi Gupta [2], Androulla Pericleous [2], Daniel E. Kadouri [2] and Robert M. Q. Shanks [1,*]

[1] Charles T. Campbell Eye Microbiology Laboratory, Department of Ophthalmology, The Eye and Ear Institute, University of Pittsburgh School of Medicine, 203 Lothrop Street, Room 1020, Pittsburgh, PA 15213, USA; romanowskieg@upmc.edu

[2] Department of Oral Biology, Rutgers School of Dental Medicine, 110 Bergen St, Newark, NJ 07103, USA; sgupta1109@gmail.com (S.G.); androulla830@gmail.com (A.P.); kadourde@sdm.rutgers.edu (D.E.K.)

* Correspondence: shanksrm@upmc.edu; Tel.: +1-(412)-647-3537

Abstract: It was previously demonstrated that predatory bacteria are able to efficiently eliminate Gram-negative pathogens including antibiotic-resistant and biofilm-associated bacteria. In this proof-of-concept study we evaluated whether two species of predatory bacteria, *Bdellovibrio bacteriovorus* and *Micavibrio aeruginosavorus*, were able to alter the survival of Gram-negative pathogens on the ocular surface. Clinical keratitis isolates of *Pseudomonas aeruginosa* (strain PAC) and *Serratia marcescens* (strain K904) were applied to the ocular surface of NZW rabbits followed by application of predatory bacteria. At time intervals, surviving pathogenic bacteria were enumerated. In addition, *B. bacteriovorus* and *S. marcescens* were applied to porcine organ culture corneas under contact lenses, and the ocular surface was examined by scanning electron microscopy. The ocular surface epithelial layer of porcine corneas exposed to *S. marcescens*, but not *B. bacteriovorus* was damaged. Using this model, neither pathogen could survive on the rabbit ocular surface for longer than 24 h. *M. aeruginosavorus* correlated with a more rapid clearance of *P. aeruginosa* but not *S. marcescens* from rabbit eyes. This study supports previous evidence that predatory bacteria are well tolerated by the cornea, but suggest that predatory bacteria do not considerably change the ability of the ocular surface to clear the tested Gram-negative bacterial pathogens from the ocular surface.

Keywords: ocular infection; predatory bacteria; *Bdellovibrio*; *Micavibrio*; *Pseudomonas aeruginosa*; *Serratia marcescens*; conjunctivitis; keratitis

1. Introduction

Predatory bacteria including *Bdellovibrio bacteriovorus* and *Micavibrio aeruginosavorus* are Gram-negative bacteria that prey upon other Gram-negative bacteria [1,2]. These species are able to prey on a wide range of antibiotic-resistant bacteria including many human pathogens [3,4] such as ocular isolates of *Pseudomonas aeruginosa* and *Serratia marcescens* [4]. *B. bacteriovorus* has a broad host-range by which it invades the bacterial cell and replicates in the bacterial periplasm, whereas *M. aeruginosavorus* exhibits a narrower host-range and acts as an epibiotic predator as it attaches to the outside of prey bacteria [5]. These predators were shown to be highly effective against bacteria in biofilms, which are notoriously recalcitrant to traditional antibiotic therapy [6–9].

We previously postulated that predatory bacteria could be used as a topical treatment for bacterial infection of the eye and demonstrated that predatory bacteria are not toxic to human ocular surface cell lines and well tolerated on the ocular surface of rabbits [4,10]. Other groups have found similar tolerability of predatory bacteria on leporine and bovine ocular surfaces [11,12]. Furthermore, intravenous and intranasal inoculation of *Micavibrio* and *Bdellovibrio* species, even at high numbers, caused no morbidity or mortality in mice, although they did mildly increase production of proinflammatory cytokines IL-6, TNF-alpha, and chemokine CXCL-1 [13], and numerous mammalian cell lines were unperturbed by *Bdellovibrio* strains [4,14,15]. Together, these data suggest that *Bdellovibrio* and *Micavibrio* species can be safely used as an experimental therapeutic. Additionally, in vivo studies had reviled that predatory bacteria have potential as "living antimicrobials" for control of pathogens. *B. bacteriovorus* have shown efficacy in limiting *Klebsiella pneumoniae* and *Yersinia pestis* proliferation in airway and systemic rodent infection models [16,17]. Similarly, they were able to prey upon *Shigella flexneri* in the hindbrain of zebrafish, promoting the survival of the zebrafish larvae [18].

Ocular infections caused by Gram-negative bacteria, such as keratitis, are associated with contact lens use and can lead to a loss of ocular acuity [19–21]. Leading causes of these infections include *Pseudomonas aeruginosa* and *Serratia marcescens* [22–24]. Antibiotic resistance has been noted among keratitis isolates and is correlated with worse clinical outcomes [24–29]. Due to the need for new approaches to treat resistant microbial infections, we evaluated the ability of *B. bacteriovorus* and *M. aeruginosavorus* to promote the clearance of keratitis isolates of *S. marcescens* and fluoroquinolone-resistant *P. aeruginosa* and from the ocular surface using a rabbit ocular surface occupancy model.

2. Results

2.1. Scanning Electron Microscopy Visualization of B. bacteriovorus 109J with Porcine Corneas Ex Vivo

As a first step in the study we visualized the interaction of predatory bacteria and the cornea in order to determine whether predatory bacteria could adhere to the corneal surface and whether there was any clear impact of this interaction using *B. bacteriovorus* 109J as a representative strain of predatory bacteria. Although a previous study demonstrated the absence of a clinical inflammatory response by rabbits, it did not evaluate the ocular surface at a microscopic level. Figure 1 depicts ex vivo porcine corneas from an organ culture model where *B. bacteriovorus* strain 109J was in contact with the ocular surface under a contact lens for 3 h. The predatory bacteria could adhere to the ocular surface, but failed to produce any clear epithelial damage similar to the mock treated (no bacteria) samples. By comparison, when using a sample Gram-negative pathogen, *S. marcescens* strain K904, under the same experimental conditions, adherent bacteria were present and were associated with erosion-like areas.

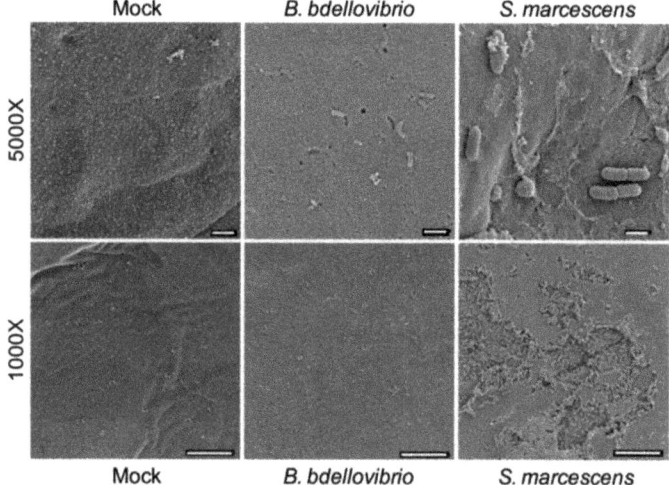

Figure 1. SEM micrographs of porcine corneal surfaces exposed to bacteria for 3 h ex vivo. Representative images are shown. Top row bars, 1 μm. Bottom row bars, 10 μm. Both *B. Bdellovibrio* strain 109J and *S. marcescens* strain K904 could adhere to the corneal surface, but *S. marcescens* was associated with damage to the epithelium.

2.2. Clearance of Fluoroquinolone Resistant P. aeruginosa But Not S. marcescens from Rabbit Ocular Surfaces Was Facilitated by Instillation of Predatory Bacteria

The survival of a fluoroquinolone-resistant keratitis isolate of *P. aeruginosa* (strain PAC) was evaluated on the ocular surface of NZW rabbits. PAC was previously shown to be susceptible to the predatory bacteria used in this study in vitro [4]. It was shown that PAC was reduced 2.13 Log_{10} CFU by *B. bacteriovorus* 109J, 3.91 Log_{10} CFU by *B. bacteriovorus* HD100 and 2.98 Log_{10} CFU by *M. aeruginosavorus* ARL-13. Predatory bacteria or saline was applied topically at 1, 3, and 5 h post-instillation of *P. aeruginosa*, and bacteria were enumerated at 0.5, 2, and 4, and 24 h (Figure 2A). No growth was measured from the samples taken at 24 h.

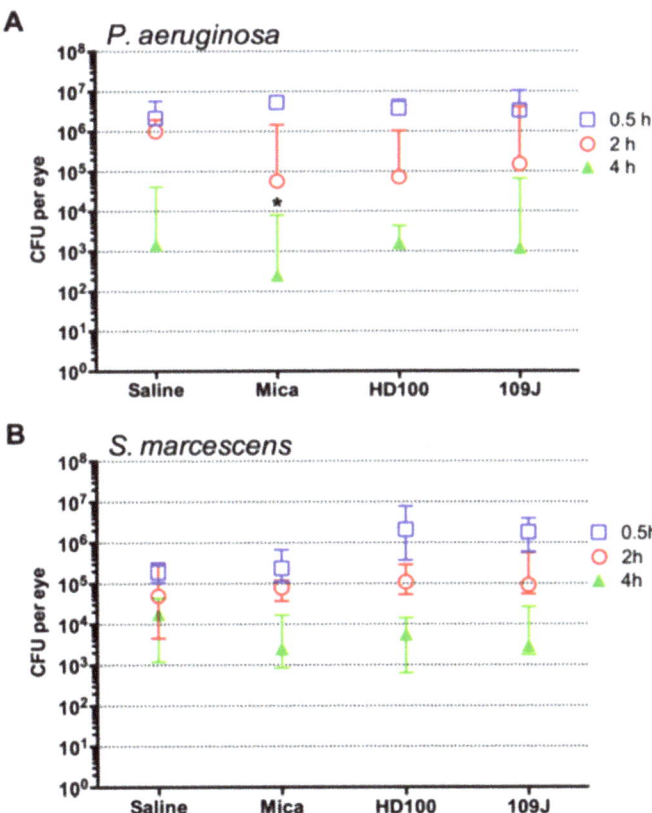

Figure 2. Predatory bacteria impact the ocular surface survival of *P. aeruginosa* and *S. marcescens*. (**A**,**B**). Medians and interquartile ranges of bacterial CFU from ocular surface of New Zealand white rabbits, (**A**). *P. aeruginosa*, n = 12 eyes per group. (**B**). *S. marcescens*, n = 8 eyes per group. Asterisks indicate significant differences from saline at the same time point ($p < 0.05$) as determined by Mann–Whitney test.

In the saline group, median PAC levels remained steady at over 10^6 CFU for the first 2 h then dropped down to just under 1500 CFU per swab at 4 h. Notably between hour 0 and 2 there was no clear reduction in PAC in the saline treated eyes, whereas the predatory bacteria treated eyes had a reduction in the number of PAC bacteria. PAC CFU dropped 2.0, 1.7, and 1.3 Log_{10} for *B. bacteriovorus* 109J, *Micavibrio* (Mica), and *B. bacteriovorus* HD100, respectively, between 2 and 4 h. PAC CFU from the Mica treatment group at 4 h post-inoculation (254 CFU) was significantly different than the saline treatment group (1483 CFU), (Mann–Whitney $p < 0.05$).

The same approach was performed with *S. marcescens* contact lens associated keratitis isolate K904, which is also susceptible to predatory bacteria in vitro [30]. Garcia et al. showed that *B. bacteriovorus* 109J reduced *S. marcescens* K904 CFU by 4.1 Log_{10} CFU but only 0.3 Log_{10} CFU by *M. aeruginosavorus* ARL-13 [30]. Here, we evaluated *S. marcescens* strain K904 predation by *B. bacteriovorus* HD100 in vitro and measured a 3.92 ± 0.17 Log_{10} CFU reduction compared to an increase of 0.17 ± 0.12 CFU change in the control samples without predatory bacteria ($n = 3$). *S. marcescens* CFU were similar at 0 and 2 h but were reduced at 4 h on the rabbit ocular surface (Figure 2B) compared with the saline control. Predatory bacteria did not significantly impact the survival of *S. marcescens* on the ocular surface.

3. Discussion

This study indicated that the predatory bacterium *B. bacteriovorus* strain 109J was not damaging to live corneas when tested in an organ culture model, which is consistent with previous studies indicating that predatory bacteria are well tolerated by tissue culture cell lines and mammals [10,12,14–16,18]. By contrast a representative ocular surface pathogen, *S. marcescens* caused clear damage to the corneal epithelia. This may be due to the many cytotoxic enzymes, such as PrtS, SlpB, and SlpE metalloproteases, and the pore-forming toxin ShlA, previously shown to be cytotoxic to ocular surface cells [31–33].

Two previous studies have evaluated the ability of predatory bacteria to reduce bacterial counts on the ocular surface; one with *Shigella flexneri* was inconclusive with respect to predation as the *S. flexneri* numbers were reduced following application of either *B. bacteriovorus* or non-pathogenic *Escherichia coli* [11]. In another study, lyophilized *B. bacteriovorus* strain 109J was used in topical treatment of calf corneas that had been infected with pathogenic *Moraxella bovis* using an infectious bovine keratoconjunctivitis model [12]. Boileau and colleagues concluded that the treatment group did not differ from the control group in treatment of the ocular surface infection [12]. By contrast, *B. bacteriovorus* treatment has been effective in reducing pathogen numbers in rat lung and zebrafish larvae hindbrain infection models [13,18] and reducing *Salmonella* numbers in the gut of chickens following oral dosing [34]. Therefore, it is clear that there are physiological limits to where predatory bacteria can be used as alternatives to antibiotics. The ocular surface is a notably hostile environment to bacteria and is considered paucibacterial with relatively few bacteria compared to other exposed mucosal surfaces [35]. Although DNA for Gram-negative bacteria have been isolated from the ocular surface following PCR amplification in several studies, it is not clear that they are constituents of the normal ocular microbiome, which is dominated by Gram-positive genera such as *Corynebacterium* and *Staphylococcus* that are resistant to the tested predatory bacteria [35–37]. Therefore, as was demonstrated for the rat gut microbiome [38], it is not expected that predatory bacteria would have a major effect on the ocular surface microbiome. Furthermore, *B. bacteriovorus* abundance has been positively correlated with a healthy gut microbiome and the absence of inflammatory disease in humans, suggesting a beneficial role for these organisms [39].

The act of swabbing or proparacaine topical anesthetic solution may have influenced the outcome of the study. Proparacaine has been shown to inhibit *Staphylococcus aureus*, but not *P. aeruginosa* growth in vitro [40]. However, a veterinary study demonstrated no significant effect of proparacaine on the number of bacteria isolated from ocular surface samples, suggesting that proparacaine did not impact this study [41]. Similarly, previous studies have demonstrated rapid clearance of *P. aeruginosa* on the ocular surface of rodents [42], suggesting that the rapid reduction of pathogen bacteria on the ocular surface was due to the innate immune system of the eye rather than due to swabbing the ocular surface.

There was a correlation of the presence of predatory bacteria with a reduction in *P. aeruginosa* CFU on the ocular surface at 2 and 4 h that only reached significance with *M. aeruginosavorus*. By contrast *S. marcescens* surface occupancy was not altered by the presence of predatory bacteria. The in vitro reduction of *S. marcescens* K904 (~4-log reduction) by *B. bacteriovorus* was higher than for *P. aeruginosa* PAC (2.1–3.9-log reduction); whereas *M. aeruginosavorus* reduced *P. aeruginosa* PAC (~3-log reduction) greater than *S. marcescens* K904 (~0.3-log reduction) in vitro [4,30]. While it is clear that these pathogens were preyed upon in vitro, whether there was active predation on the ocular surface was not formally determined in this pilot study. Indeed, predatory bacteria may stimulate the immune system to promote clearance of *P. aeruginosa*. Consistent with this hypothesis, in a zebrafish infection study, predatory bacteria preyed upon *S. flexneri* in the hind brain, but full clearance of the pathogen required both the predatory bacteria and the immune system [18]. On the ocular surface, colonization of *Corynebacterium* species can promote resistance to *Pseudomonas* infections that is dependent upon an IL-17 signaling mechanism [43]. It is possible that the predatory bacteria are invoking a similar protective immune response in rabbits.

Together, these data suggest that predatory bacteria are not damaging to the corneal epithelium and can influence the occupancy of pathogens on the ocular surface, but that they are not an effective method of clearing pathogens beyond that of the natural host defense systems.

4. Materials and Methods

4.1. Bacterial Strains and Culture

Keratitis isolates of *P. aeruginosa* strain PAC [44] and *S. marcescens* strain K904 [45] were used in this study. The *P. aeruginosa* strain was determined to be resistant to fluoroquinolone antibiotics (ciprofloxacin, gatifloxacin, ofloxacin, levofloxacin, and moxifloxacin) in a College of American Pathologists (CAP) and Clinical Laboratory Improvement Amendments (CLIA) certified microbiology laboratory following Clinical and Laboratory Standards Institute (CLSI) guidelines [44,46]. Susceptibility was interpreted using the CLSI (Clinical & Laboratory Standards Institute) serum standards and procedures for disk diffusion [46], and later determined using E-tests [47].

These bacteria were maintained in glycerol frozen stocks and were streaked to single colonies on TSA medium with 5% red blood cells (Blood agar) (Remel, Lenexa, KS) before use as described below. Bacteria were also cultured with lysogeny broth (LB) and LB with agar [48]. The predatory bacteria used in the study were *B. bacteriovorus* 109J [49] *B. bacteriovorus* HD100 (ATCC 15356) [50], and *M. aeruginosavorous* strain ARL-13 [51]. Predator lysates (cocultures) were prepared as reported previously [14,16]. In brief, *B. bacteriovorus* and *M. aeruginosavorus* were incubated with *E. coli* strain WM3064 (1×10^9 CFU/mL) at 30°C for 24 and 72 h, respectively. The cleared lysates were filtered several times through a 0.45-μm Millex®-HV pore-size filter (Millipore, Billerica, MA, USA) in order to remove residual prey. Predators were washed and concentrated by sequential centrifugation cycles. The final predator pellets were re-suspended in Phosphate Buffered Saline (PBS) to reach final concentrations of 1×10^{10} PFU/mL *B. bacteriovorus* and 1×10^9 PFU/mL *M. aeruginosavorus*.

4.2. Scanning Electron Microscopy

B. bacteriovorus strain 109J, prepared in PBS as described above, were applied in 50 μL samples to the surfaces of ex vivo porcine corneas and contact lenses (CL) were applied. PBS alone was used as a negative control, and *S. marcescens* strain K904 in PBS (3×10^9) was applied as a control pathogen. Porcine eyes were obtained from Sierra Medical (Whittier, CA, USA) and corneal organ culture was performed as previously described but without antibiotics [52,53]. The ex vivo corneas were incubated at 37 °C for 3 h, then the CL were removed. The corneas were rinsed twice with PBS to remove non- or loosely adherent bacteria and fixed with glutaraldehyde (3%) overnight at room temperature. Corneas were then washed with PBS and post-fixed using aqueous osmium tetroxide (1%), dehydrated using increasing ethanol concentrations (30–100%), immersed in hexamethyldisilazane, air dried, and sputter coated with gold/palladium (6 nm). A

JEOL JSM-6335F scanning electron microscope at 3 kV with the secondary electron imaging detector was used for imaging.

4.3. In Vitro Predation Assay

Susceptibility of *S. marcescens* strain K904 to *B. bacteriovorus* strain HD100 was tested as previously described [30]. HD100 and K904 were combined in 14 mL Falcon™ round-bottom polypropylene tubes by adding 0.4 mL of harvested predators (5×10^8 PFU/mL) to 0.4 washed *S. marcescens* (4×10^9 CFU/mL) and 1.2 mL HEPES buffer (HEPES at 25 mM supplemented with $CaCl_2$ at 2 mM and $MgCl_2$ at 3 mM). These were incubated at 30 °C on a rotary shaker set at 30 rpm. A control without *B. bacteriovorus* was included as a control. Colony forming units of *S. marcescens* were determined by dilution plating on LB agar plates after 24 of coculture. The experiment was repeated three times.

4.4. Rabbit Ocular Surface Occupancy Model

This study conformed to the ARVO Statement on the Use of Animals in Ophthalmic and Vision Research and was approved by the University of Pittsburgh Institutional Animal Care and Use Committee (Protocol 15025331). Female New Zealand white rabbits weighing 1.1–1.4 kg, were obtained from Charles River Oakwood rabbitry.

For the inocula, *P. aeruginosa* strain PAC and *S. marcescens* strain K904 were swabbed onto 5 blood agar plates and incubated at 37 °C overnight. Bacteria were scraped off the plates using a cotton tipped applicator and suspended in 5 mL of phosphate buffered saline (PBS) and adjusted to a culture density of OD_{600} of 5 in PBS. *P. aeruginosa* and *S. marcescens* inocula colony counts were determined using the EddyJet 2 spiral plating system (Neutec Group Inc., Farmingdale, NY, USA) on blood agar plates. The plates were incubated overnight at 37 °C for *P. aeruginosa* and 30 °C for *S. marcescens* and resulting colonies were enumerated using the automated Flash and Grow colony counting system (Neutec Group), with (~5×10^8 CFU) in 50 µL samples of bacteria that were applied to the ocular surface of both eyes of unanesthetized rabbits. Fifty µL of the predatory bacteria were installed into the rabbits' eyes and consisted of 2×10^8 PFU/mL for *B. bacteriovorus* and 2×10^7 PFU/mL for *M. aeruginosavorus*.

The ocular surfaces of both rabbit eyes were inoculated with *P. aeruginosa* ($n = 12$ rabbits) and *S. marcescens* ($n = 8$ rabbits). At 0.5, 2, 4, and 24 post-inoculation, each eye was cultured following topical anesthesia with 2 drops of 0.5% proparacaine (Proparacaine Hydrochloride Ophthalmic Solution, USP, 0.5%, Sandoz Inc., Princeton, NJ, USA) by inserting a Dacron-tipped applicator into the upper and lower fornices and gently manipulating the swab over the conjunctival and corneal surfaces. Swabs were placed into 1 mL of PBS and kept on ice. Dilutions (1:100 and 1:10,000) of the samples were made in PBS. The undiluted and diluted samples were plated on blood agar plates to enumerate bacteria as describe above. At 1, 3, and 5 h post-inoculation, 50 µL topical drops with predatory bacteria or saline were applied to eyes. *P. aeruginosa* and *S. marcescens* remaining of the ocular surface were enumerated as described above. Median colony forming units (CFU) were compared using non-parametric analysis with GraphPad Prism software.

4.5. Statistical Analysis

Mann–Whitney analysis was performed using GraphPad Prism statistical software version 6.0. p-values less than 0.05 were considered significant.

Author Contributions: E.G.R. contributed to the research design; E.G.R. obtained Institutional Animal Care and Use Committee approval and upheld regulations; E.G.R. directed rabbit experiments; and E.G.R. coauthored and reviewed the final manuscript. S.G. contributed to the research design; S.G. directed preparation of predatory bacteria; S.G. determined predatory bacteria numbers; and S.G. reviewed the final manuscript. A.P. prepared predatory bacteria; A.P. determined predatory bacteria numbers, and the review and writing of the final manuscript. D.E.K. contributed to the research design; D.E.K. participated in the writing and review of the final manuscript. R.M.Q.S. contributed to the research design; R.M.Q.S. prepared bacterial inocula; R.M.Q.S. participated in the writing and review of the final manuscript. All authors have read and agreed to the published version of the manuscript.

Funding: The Ophthalmology Department has received support from: NIH grants P30-EY08098, the Eye and Ear Foundation of Pittsburgh, Research to Prevent Blindness. This work was funded by the U.S. Army Research Office and the Defense Advanced Research Projects Agency (DARPA) and was accomplished under Cooperative Agreement Number W911NF-15-2-0036 to D.E.K. and R.M.Q.S. The views and conclusions contained in this document are those of the authors and should not be interpreted as representing the official policies, either expressed or implied, of the Army Research Office, DARPA, or the U.S. Government. The U.S. Government is authorized to reproduce and distribute reprints for Government purposes notwithstanding any copyright notation hereon.

Institutional Review Board Statement: Animal studies conformed to the Association for Research in Vision and Ophthalmology Statement on the Use of Animals in Ophthalmic and Vision Research, and was approved by the University of Pittsburgh Institutional Animal Care and Use Committee (Protocol 15025331).

Informed Consent Statement: Not applicable.

Data Availability Statement: Data are available on request.

Acknowledgments: The authors thank Kimberly Brothers, Nicholas Stella, and Kathleen Yates for expert technical help and Jonathan Franks at the Center for Biological Imaging at the University of Pittsburgh for SEM analysis.

Conflicts of Interest: The authors declare no conflict of interest.

References

1. Dwidar, M.; Monnappa, A.K.; Mitchell, R.J. The dual probiotic and antibiotic nature of Bdellovibrio bacteriovorus. *BMB Rep.* **2012**, *45*, 71–78. [CrossRef]
2. Sockett, R.E. Predatory lifestyle of *Bdellovibrio bacteriovorus*. *Annu. Rev. Microbiol.* **2009**, *63*, 523–539. [CrossRef]
3. Kadouri, D.E.; To, K.; Shanks, R.M.; Doi, Y. Predatory bacteria: A potential ally against multidrug-resistant Gram-negative pathogens. *PLoS ONE* **2013**, *8*, e63397. [CrossRef]
4. Shanks, R.M.; Davra, V.R.; Romanowski, E.G.; Brothers, K.M.; Stella, N.A.; Godboley, D.; Kadouri, D.E. An Eye to a Kill: Using Predatory Bacteria to Control Gram-Negative Pathogens Associated with Ocular Infections. *PLoS ONE* **2013**, *8*, e66723. [CrossRef]
5. Pasternak, Z.; Njagi, M.; Shani, Y.; Chanyi, R.; Rotem, O.; Lurie-Weinberger, M.N.; Koval, S.; Pietrokovski, S.; Gophna, U.; Jurkevitch, E. In and out: An analysis of epibiotic vs periplasmic bacterial predators. *ISME J.* **2014**, *8*, 625–635. [CrossRef]
6. Kadouri, D.; O'Toole, G.A. Susceptibility of biofilms to *Bdellovibrio bacteriovorus* attack. *Appl. Environ. Microbiol.* **2005**, *71*, 4044–4051. [CrossRef] [PubMed]
7. Kadouri, D.; Venzon, N.C.; O'Toole, G.A. Vulnerability of pathogenic biofilms to *Micavibrio aeruginosavorus*. *Appl. Environ. Microbiol.* **2007**, *73*, 605–614. [CrossRef]
8. Dharani, S.; Kim, D.H.; Shanks, R.M.Q.; Doi, Y.; Kadouri, D.E. Susceptibility of colistin-resistant pathogens to predatory bacteria. *Res. Microbiol.* **2018**, *169*, 52–55. [CrossRef] [PubMed]
9. Duncan, M.C.; Gillette, R.K.; Maglasang, M.A.; Corn, E.A.; Tai, A.K.; Lazinski, D.W.; Shanks, R.M.Q.; Kadouri, D.E.; Camilli, A. High-throughput analysis of gene function in the bacterial predator *Bdellovibrio bacteriovorus*. *mBio* **2019**, *10*, e01040-19. [CrossRef]
10. Romanowski, E.G.; Stella, N.A.; Brothers, K.M.; Yates, K.A.; Funderburgh, M.L.; Funderburgh, J.L.; Gupta, S.; Dharani, S.; Kadouri, D.E.; Shanks, R.M. Predatory bacteria are nontoxic to the rabbit ocular surface. *Sci. Rep.* **2016**, *6*, 30987. [CrossRef]
11. Nakamura, M. Alteration of *Shigella* pathogenicity by other bacteria. *Am. J. Clin. Nutr.* **1972**, *25*, 1441–1451. [CrossRef]
12. Boileau, M.J.; Mani, R.; Breshears, M.A.; Gilmour, M.; Taylor, J.D.; Clickenbeard, K.D. Efficacy of *Bdellovibrio bacteriovorus* 109J for the treatment of dairy calves with experimentally induced infectious bovine keratoconjunctivitis. *Am. J. Vet. Res.* **2016**, *77*, 1017–1028. [CrossRef]
13. Shatzkes, K.; Chae, R.; Tang, C.; Ramirez, G.C.; Mukherjee, S.; Tsenova, L.; Connell, N.D.; Kadouri, D.E. Examining the safety of respiratory and intravenous inoculation of *Bdellovibrio bacteriovorus* and *Micavibrio aeruginosavorus* in a mouse model. *Sci. Rep.* **2015**, *5*, 12899. [CrossRef] [PubMed]
14. Gupta, S.; Tang, C.; Tran, M.; Kadouri, D.E. Effect of predatory bacterio on human cell lines. *PLoS ONE* **2016**, *11*, e0161242. [CrossRef]
15. Monnappa, A.K.; Bari, W.; Choi, S.Y.; Mitchell, R.J. Investigating the responses of human epithelial cells to predatory bacteria. *Sci. Rep.* **2016**, *15*, 33485. [CrossRef]
16. Shatzkes, K.; Singleton, E.; Tang, C.; Zuena, M.; Shukla, S.; Gupta, S.; Dharani, S.; Onyile, O.; Rinaggio, J.; Connell, N.D.; et al. Predatory bacteria attenuate *Klebsiella pneumoniae* burden in rat lungs. *mBio* **2016**, *7*, e01847-16. [CrossRef]
17. Findlay, J.S.; Flick-Smith, H.C.; Keyser, E.; Cooper, I.A.; Williamson, E.D.; Oyston, P.C.F. Predatory bacteria can protectx SKH-1 mice from a lethal plague challenge. *Sci. Rep.* **2019**, *9*, 7525. [CrossRef]

18. Willis, R.A.; Moore, C.; Mazon-Moya, M.; Krokowski, S.; Lambert, C.; Till, R.; Mostowy, S.; Sockett, R.E. Injections of predatory bacteria work alongside host immune cells to treat *Shigella* infections in zebrafish larvae. *Curr. Biol.* **2016**, *26*, 3343–3351. [CrossRef] [PubMed]
19. Das, S.; Sheorey, H.; Taylor, H.R.; Vajpayee, R.B. Association between cultures of contact lens and corneal scraping in contact lens related microbial keratitis. *Arch. Ophthalmol.* **2007**, *125*, 1182–1185. [CrossRef]
20. Green, M.; Sara, S.; Hughes, I.; Apel, A.; Stapleton, F. Trends in contact lens microbial keratitis 1999 to 2015: A retrospective clinical review. *Clin. Exp. Ophthalmol.* **2019**, *47*, 726–732. [CrossRef]
21. Mah-Sadorra, J.H.; Najjar, D.M.; Rapuano, C.J.; Laibson, P.R.; Cohen, E.J. *Serratia* corneal ulcers: A retrospective clinical study. *Cornea* **2005**, *24*, 793–800. [CrossRef]
22. Lakhundi, S.; Siddiqui, R.; Khan, N.A. Pathogenesis of microbial keratitis. *Microb. Pathog.* **2017**, *104*, 97–109. [CrossRef]
23. Hume, E.B.; Willcox, M.D. Emergence of Serratia marcescens as an ocular surface pathogen. *Arch. Soc. Esp. Oftalmol.* **2004**, *79*, 475–477.
24. Hilliam, Y.; Kaye, S.; Winstanley, C. *Pseudomonas aeruginosa* and microbial keratitis. *J. Med. Microbiol.* **2020**, *69*, 3–13. [CrossRef] [PubMed]
25. Green, M.; Apel, A.; Stapleton, F. Risk factors and causative organisms in microbial keratitis. *Cornea* **2008**, *27*, 22–27. [CrossRef] [PubMed]
26. Willcox, M.D. Review of resistance of ocular isolates of Pseudomonas aeruginosa and staphylococci from keratitis to ciprofloxacin, gentamicin and cephalosporins. *Clin. Exp. Optom.* **2011**, *94*, 161–168. [CrossRef]
27. Shen, E.P.; Hsieh, Y.T.; Chu, H.S.; Chang, S.C.; Hu, F.R. Correlation of *Pseudomonas aeruginosa* genotype with antibiotic susceptibility and clinical features of induced central keratitis. *Investig. Ophthalmol. Vis. Sci.* **2014**, *56*, 365–371. [CrossRef] [PubMed]
28. Vazirani, J.; Wurity, S.; Ali, M.H. Multidrug-Resistant *Pseudomonas aeruginosa* Keratitis: Risk Factors, Clinical Characteristics, and Outcomes. *Ophthalmology* **2015**, *122*, 2110–2114. [CrossRef]
29. Fernandes, M.; Vira, D.; Medikonda, R.; Kumar, N. Extensively and pan-drug resistant *Pseudomonas aeruginosa* keratitis: Clinical features, risk factors, and outcome. *Graefes Arch. Clin. Exp. Ophthalmol.* **2016**, *254*, 315–322. [CrossRef]
30. Garcia, C.J.; Pericleous, A.; Elsayed, M.; Tran, M.; Gupta, S.; Callaghan, J.D.; Stella, N.A.; Franks, J.M.; Thibodeau, P.H.; Shanks, R.M.Q.; et al. Serralysin family metalloproteases protects *Serratia marcescens* from predation by the predatory bacteria *Micavibrio aeruginosavorus*. *Sci. Rep.* **2018**, *8*, 14025. [CrossRef]
31. Shanks, R.M.; Stella, N.A.; Hunt, K.M.; Brothers, K.M.; Zhang, L.; Thibodeau, P.H. Identification of SlpB, a cytotoxic protease from *Serratia marcescens*. *Infect. Immun.* **2015**, *83*, 2907–2916. [CrossRef]
32. Stella, N.A.; Callaghan, J.D.; Zhang, L.; Brothers, K.M.; Kowalski, R.P.; Huang, J.J.; Thibodeau, P.H.; Shanks, R.M.Q. SlpE is a calcium-dependent cytotoxic metalloprotease associated with clinical isolates of *Serratia marcescens*. *Res. Microbiol.* **2017**, *168*, 567–574. [CrossRef] [PubMed]
33. Stella, N.A.; Brothers, K.M.; Shanks, R.M.Q. Differential susceptibility of airway and ocular surface cell lines to FlhDC-mediated virulence factors PhlA and ShlA from *Serratia marcescens*. *J. Med. Microbiol.* **2021**, *70*, 001292. [CrossRef]
34. Atterbury, R.J.; Hobley, L.; Till, R.; Lambert, C.; Capeness, M.J.; Lerner, T.R.; Fenton, A.K.; Barrow, P.; Sockett, R.E. Effects of orally administered *Bdellovibrio bacteriovorus* on the well-being and *Salmonella* colonization of young chicks. *Appl. Environ. Microbiol.* **2011**, *77*, 5794–5803. [CrossRef]
35. Doan, T.; Akileswaran, L.; Andersen, D.; Johnson, B.; Ko, N.; Shrestha, A.; Shestopalov, V.; Lee, C.S.; Lee, A.Y.; Van Gelder, R.D. Paucibacterial microbiome and resident DNA virome of the healthy conjunctiva. *Investig. Ophthalmol. Vis. Sci.* **2016**, *57*, 5116–5126. [CrossRef]
36. Ozkan, J.; Willcox, M.D. The ocular microbiome: Molecular characterization of a unique and low microbial environment. *Curr. Eye Res.* **2019**, *44*, 685–694. [CrossRef] [PubMed]
37. Petrillo, F.; Pignataro, D.; Lavano, M.A.; Santella, B.; Folliero, V.; Zannella, C.; Astarita, C.; Gagliano, C.; Franci, G.; Avitabile, T.; et al. Current evidence on the ocular surface microbiota and related diseases. *Microorganisms* **2020**, *8*, 1033. [CrossRef]
38. Shatzkes, K.; Tang, C.; Singleton, E.; Shukla, S.; Zuena, M.; Gupta, S.; Dharani, S.; Rinaggio, J.; Connell, N.D.; Kadouri, D.E. Effect of predatory bacteria on the gut bacterial microbiota in rats. *Sci. Rep.* **2017**, *7*, 43483. [CrossRef]
39. Lebba, V.; Santangelo, F.; Totino, V.; Nicoletti, M.; Gagliardi, A.; De Biase, R.V.; Cucchiara, S.; Nencioni, L.; Conte, M.P.; Schippa, S. Higher prevalence and abundance of *Bdellovibrio bacteriovorus* in the human gut of healthy subjects. *PLoS ONE* **2013**, *8*, e61608. [CrossRef]
40. Dantas, P.E.; Uesugui, E.; Nishiwaki-Dantas, M.C.; Mimica, L.J. Antibacterial activity of anaesthetic solutions and preservatives: An in vitro comparative study. *Cornea* **2000**, *19*, 353–354. [CrossRef]
41. Edwards, S.G.; Maggs, D.J.; Byrne, B.A.; Kass, P.H.; Lassaline, M.E. Effect of topical application of 0.5% proparacaine on corneal culture results from 33 dogs, 12 cats, and 19 horses with spontaneously arising ulcerative keratitis. *Vet. Ophthalmol.* **2019**, *22*, 415–422. [CrossRef]
42. Mun, J.J.; Tam, C.; Kowbel, D.; Hawgood, S.; Barnett, M.J.; Evans, D.J.; Fleiszig, S.M. Clearance of *Pseudomonas aeruginosa* from a healthy ocular surface involves surfactant protein D and is compromised by bacterial elastase in a murine null-infection model. *Infect. Immun.* **2009**, *77*, 2392–2398. [CrossRef]

43. St Leger, A.J.; Desai, J.V.; Drummond, R.A.; Kugadas, A.; Almaghrabi, F.; Silver, P.; Raychaudhuri, K.; Gadjeva, M.; Iwakura, Y.; Lionakis, M.S.; et al. An ocular commensal protects against corneal infection by driving an interleukin-17 response from mucosal γδ T cells. *Immunity* **2017**, *47*, 148–158. [CrossRef]
44. Kowalski, R.P.; Pandya, A.N.; Karenchak, L.M.; Romanowski, E.G.; Husted, R.C.; Ritterband, D.C.; Shah, M.K.; Gordon, Y.J. An in vitro resistance study of levofloxacin, ciprofloxacin, and ofloxacin using keratitis isolates of *Staphylococcus aureus* and *Pseudomonas aeruginosa*. *Ophthalmology* **2001**, *108*, 1826–1829. [CrossRef]
45. Kalivoda, E.J.; Stella, N.A.; Aston, M.A.; Fender, J.E.; Thompson, P.P.; Kowalski, R.P.; Shanks, R.M. Cyclic AMP negatively regulates prodigiosin production by *Serratia marcescens*. *Res. Microbiol.* **2010**, *161*, 158–167. [CrossRef]
46. Cockerill, F.R. *Performance Standards for Antimicrobial Susceptibility Testing*; Twenty-Third Informational Supplement; CLSI: Wayne, PA, USA, 2013.
47. Kowalski, R.P.; Dhaliwal, D.K.; Karenchak, L.M.; Romanowski, E.G.; Mah, F.S.; Ritterband, D.C.; Gordon, Y.J. Gatifloxacin and moxifloxacin: An in vitro susceptibility comparison to levofloxacin, ciprofloxacin, and ofloxacin using bacterial keratitis isolates. *Am. J. Ophthalmol.* **2003**, *136*, 500–505. [CrossRef]
48. Bertani, G. Studies on lysogenesis. I. The mode of phage liberation by lysogenic Escherichia coli. *J. Bacteriol.* **1951**, *62*, 293–300. [CrossRef] [PubMed]
49. Rittenberger, S.C. Nonidentity of *Bdellovibrio bacteriovorus* strains 109D and 109J. *J. Bacteriol.* **1972**, *109*, 432–433. [CrossRef] [PubMed]
50. Stolp, H.; Starr, M.P. Bdellovibrio bacteriovorus Gen. Et Sp. N., A predatory, ectoparasitic, and bacteriolytic microorganism. *Antonie Van Leeuwenhoek* **1963**, *29*, 217–248. [CrossRef]
51. Lambina, V.A.; Afinogenova, A.V.; Romay Penobad, Z.; Konovalova, S.M.; Andreev, L.V. New species of exoparasitic bacteria of the genus Micavibrio infecting gram-negative bacteria. *Mikrobiologiia* **1983**, *52*, 777–780.
52. Brothers, K.M.; Stella, N.A.; Hunt, K.M.; Romanowski, E.G.; Liu, X.; Klarlund, J.K.; Shanks, R.M. Putting on the brakes: Bacterial impediment of wound healing. *Sci. Rep.* **2015**, *5*, 14003. [CrossRef]
53. Xu, K.P.; Li, X.F.; Yu, F.S. Corneal organ culture model for assessing epithelial responses to surfactants. *Toxicol. Sci.* **2000**, *58*, 306–314. [CrossRef]

Article

Enhancement of Antibiofilm Activity of Ciprofloxacin against *Staphylococcus aureus* by Administration of Antimicrobial Peptides

Muhammad Yasir [1,*], Debarun Dutta [1,2] and Mark D. P. Willcox [1]

1 School of Optometry and Vision Science, University of New South Wales, Sydney 2052, Australia; d.dutta@aston.ac.uk (D.D.); m.willcox@unsw.edu.au (M.D.P.W.)
2 Optometry and Vision Science Research Group, Optometry School, University of Aston, Birmingham B4 7ET, UK
* Correspondence: m.yasir@unsw.edu.au; Tel.: +61-414-941-761

Abstract: *Staphylococcus aureus* can develop resistance by mutation, transfection or biofilm formation. Resistance was induced in *S. aureus* by growth in sub-inhibitory concentrations of ciprofloxacin for 30 days. The ability of the antimicrobials to disrupt biofilms was determined using crystal violet and live/dead staining. Effects on the cell membranes of biofilm cells were evaluated by measuring release of dyes and ATP, and nucleic acids. None of the strains developed resistance to AMPs while only *S. aureus* ATCC 25923 developed resistance (128 times) to ciprofloxacin after 30 passages. Only peptides reduced biofilms of ciprofloxacin-resistant cells. The antibiofilm effect of melimine with ciprofloxacin was more (27%) than with melimine alone at 1X MIC ($p < 0.001$). Similarly, at 1X MIC the combination of Mel4 and ciprofloxacin produced more (48%) biofilm disruption than Mel4 alone ($p < 0.001$). Combinations of either of the peptides with ciprofloxacin at 2X MIC released ≥ 66 nM ATP, more than either peptide alone ($p \leq 0.005$). At 2X MIC, only melimine in combination with ciprofloxacin released DNA/RNA which was three times more than that released by melimine alone ($p = 0.043$). These results suggest the potential use of melimine and Mel4 with conventional antibiotics for the treatment of *S. aureus* biofilms.

Keywords: *Staphylococcus aureus*; antibiotic resistance; biofilms; antimicrobial peptides; ciprofloxacin; combined effect

Citation: Yasir, M.; Dutta, D.; Willcox, M.D.P. Enhancement of Antibiofilm Activity of Ciprofloxacin against *Staphylococcus aureus* by Administration of Antimicrobial Peptides. *Antibiotics* **2021**, *10*, 1159. https://doi.org/10.3390/antibiotics10101159

Academic Editor: Jean-Marc Sabatier

Received: 21 August 2021
Accepted: 21 September 2021
Published: 24 September 2021

Publisher's Note: MDPI stays neutral with regard to jurisdictional claims in published maps and institutional affiliations.

Copyright: © 2021 by the authors. Licensee MDPI, Basel, Switzerland. This article is an open access article distributed under the terms and conditions of the Creative Commons Attribution (CC BY) license (https://creativecommons.org/licenses/by/4.0/).

1. Introduction

Staphylococcus aureus is a major human pathogen that can cause several recalcitrant infections (deep-seated abscess, osteomyelitis, and endocarditis) due to the acquisition of antibiotic resistance and formation of biofilm on living tissues and medical devices [1,2]. Methicillin-resistant *S. aureus* (MRSA) has been named as a "serious threat" by the Center for Disease Control and Prevention [3,4]. Approximately 11,000 people die each year from a MRSA-related infection in the United States alone [5,6]. So far, there are limited reports on antimicrobial compounds that are able to control biofilm-associated infections caused by *S. aureus* [7].

Various strategies such as physical removal of materials colonized with bacteria or delivery of high doses of antibiotics at the site of infections have been used to treat biofilm-associated infection [8]. However, due to poor penetration of antibiotics through the extracellular polysaccharide matrix of biofilms and survival of biofilm-embedded cells, even the use of high levels of antibiotics can result in low cure rates for infections [9]. Moreover, high doses of antibiotics may cause cytotoxicity to human cells. Therefore, combinations of different antimicrobials may be required [10].

Several antimicrobial peptides (AMPs) are known to have strong antibiofilm activity against bacterial biofilms [11–13]. They can prevent bacterial attachment to surfaces (a first step toward biofilm formation) and destroy already developed biofilms by causing detachment or killing of biofilm-embedded cells [11,13,14]. They can also enhance the activity of antibiotics against biofilms when used in combination [13,15–17]. These combined treatments may become an important part of treating biofilm-related infections, such as chronic wounds or biomaterial-associated infections caused by *S. aureus* [18]. In combination treatments, one mode of action that has been proposed is that the antibiotics bind to teichoic acids of staphylococcal cell wall which reduces the interaction with AMPs and facilitates their interaction with bacterial membranes. In this way, AMPs act on the cell membranes and antibiotics target cell wall and/or inhibit biosynthesis of nucleic acids and proteins [19,20].

Melimine (TLISWIKNKRKQRPRVSRRRRRRGGRRRR) and Mel4 (KNKRKRRRR RRGGRRRR) are cationic AMPs which have a wide spectrum of activity targeting clinical isolates of Gram-negative and Gram-positive bacteria (including MRSA and multidrug-resistant *P. aeruginosa*), fungi and protozoa such as *Acanthamoeba* [21,22]. Both AMPs are non-cytotoxic at well above active concentrations [21,22]. Melimine causes hemolysis of horse red blood cells at concentrations 15 times higher than its minimum inhibitory concentration (MIC) [23] while Mel4 causes < 5% hemolysis even at concentrations 17 times higher than its MIC [23]. Melimine and Mel4 can synergize with ciprofloxacin against planktonic as well as biofilm forms of *P. aeruginosa* [24]. Ciprofloxacin is a broad-spectrum antibiotic, active against both Gram-positive and Gram-negative bacteria. Ciprofloxacin kills bacteria by binding to bacterial enzymes DNA gyrase and topoisomerase IV. After binding, the enzyme undergoes conformational changes and breaks the DNA, and ciprofloxacin prevents religation of the broken DNA which ultimately stops DNA replication [25]. Both AMPs in combination with ciprofloxacin destroy *P. aeruginosa* biofilms at concentrations lower than their MICs [13]. Both AMPs act on the cell membranes of planktonic cells of *P. aeruginosa* and this results in release of cellular contents [13]. However, it is not known whether peptides alone or in combination with antibiotics are active against *S. aureus* biofilms or can act in a similar way as they do to *P. aeruginosa* biofilms. The current study investigates the interaction of AMPs melimine or Me4 alone or in combination with ciprofloxacin against *S. aureus* biofilm in conjunction with their mode of activity.

2. Results

2.1. Minimal Inhibitory Concentration and Minimal Bactericidal Concentration

Table 1 represents the MICs and MBCs values of both the peptides and ciprofloxacin. Melimine and Mel4 had the lowest MICs of 62.5 µg/mL and 125 µg/mL, respectively, against *S. aureus* ATCC 6538. For all other strains, there were slightly higher MICs, 125 µg/mL for melimine and 250 µg/mL for Mel4, except for *S. aureus* ATCC 25923 for which Mel4 had the highest MIC value of 500 µg/mL (Table 1). Ciprofloxacin had similar MICs (0.5 µg/mL) and MBCs (1 µg/mL) against all the tested strains except for *S. aureus* ATCC 6538 for which ciprofloxacin had the same MIC and MBC values of 0.5 µg/mL (Table 1).

2.2. Development of Resistance to AMPs and Ciprofloxacin

The growth curves of *S. aureus* ATCC 25923 at sub-MICs of melimine, Mel4 or ciprofloxacin over 24 h are presented in Figure 1. The growth of *S. aureus* ATCC 25923 at its sub-MIC for ciprofloxacin was similar to growth without the antimicrobial. Melimine and Mel4 affected the growth rate of *S. aureus* after 6 h. Exposure to melimine resulted in slightly less growth than exposure to Mel4 over 24 h.

Table 1. MIC and MBC values of melimine, Mel4 and ciprofloxacin against *S. aureus*.

Bacterial Strains	Melimine		Mel4		Ciprofloxacin	
	MIC μM (μg·mL^{-1})	MBC μM (μg·mL^{-1})	MIC μM (μg·mL^{-1})	MBC μM (μg·mL^{-1})	MIC μM (μg·mL^{-1})	MBC μM (μg·mL^{-1})
S. aureus 31	33.01 (125)	66.02 (250)	106.48 (250)	212.96 (500)	1.50 (0.5)	3.01 (1)
S. aureus 38	33.01 (125)	66.02 (250)	106.48 (250)	212.96 (500)	1.50 (0.5)	3.01 (1)
S. aureus ATCC 6538	16.50 (62.5)	16.50 (62.5)	53.24 (125)	53.24 (125)	1.50 (0.5)	1.50 (0.5)
S. aureus ATCC 25923	33.01 (125)	66.02 (250)	212.96 (500)	212.96 (500)	1.50 (0.5)	3.01 (1)

MBC = minimum bactericidal concentration that kills ≥ 99.99% of bacteria of bacterial population compared to positive control; MIC = minimum inhibitory concentration that kills ≥ 90% of bacterial population when compared to the positive control.

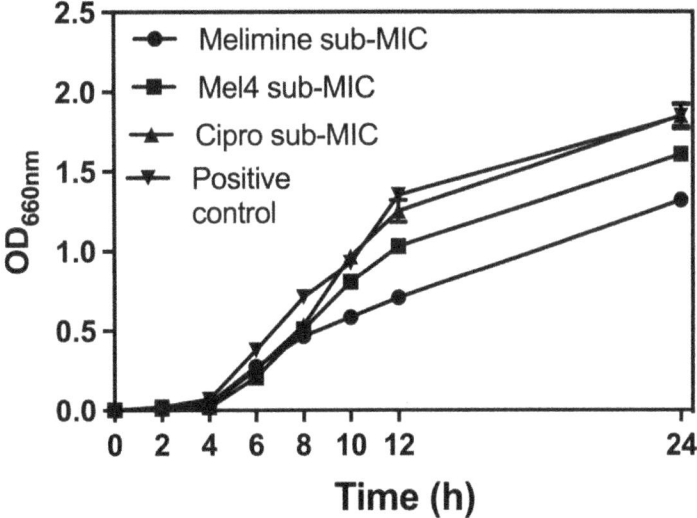

Figure 1. Growth curves for *S. aureus* ATCC 25923 at sub-MIC of the antimicrobial peptides (AMPs) melimine and Mel4 or ciprofloxacin (Cipro). Melimine and Mel4 reduced the overall bacterial growth over 24 h of experiments while ciprofloxacin and the positive control (without any antimicrobial) had similar growth characteristics after 24 h experiment.

Of all the tested strains, only *S. aureus* ATCC 25923 was able to develop resistance to ciprofloxacin. Changes in MICs of *S. aureus* ATCC 25923 after exposure to sub-MICs of melimine, Mel4 or ciprofloxacin over 30 days are presented in Figure 2. The MICs of melimine and Mel4 did not change over time, suggesting a limited potential of resistance development to these peptides. Compared to the peptides, there was rapid development of resistance to ciprofloxacin. Resistance developed to ciprofloxacin after 7 days of serial passage with an initial 4-fold increase in MIC. The MIC increased 64-fold after 15 passages and 128-fold by 30 passages (Figure 2).

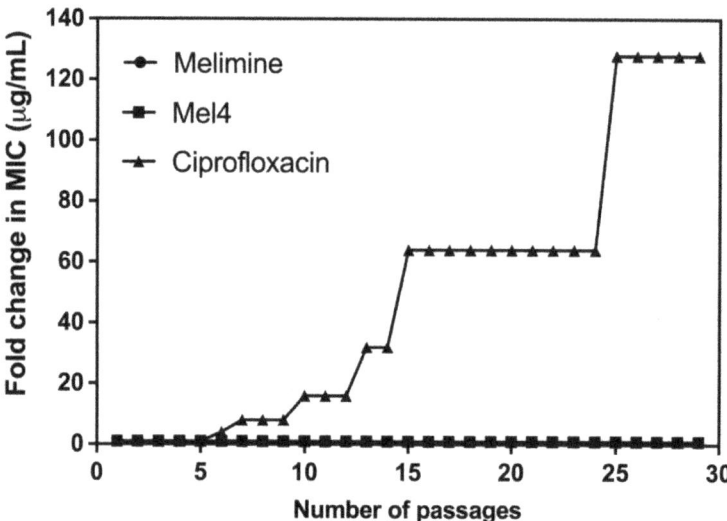

Figure 2. Increase in MIC values of ciprofloxacin, melimine or Mel4 against *S. aureus* ATCC 25923 after exposing bacteria at their sub-MIC for 30 consecutive days. The MIC values of melimine and Mel4 did not change over time and overlap at the bottom of the figure.

2.3. Inhibition of Biofilm Formation by AMPs and Ciprofloxacin Alone or in Combination

Ciprofloxacin did not inhibit the biofilm formation of the ciprofloxacin-resistant cells of *S. aureus* ATCC 25923 at any concentration tested ($p > 0.999$; Figure 3A). Melimine and Mel4 inhibited biofilm formation at 0.5X MIC by 82% and 78%, respectively, compared to the negative control ($p < 0.001$). There was similar biofilm inhibition with both the AMPs at 0.5X MIC ($p > 0.999$). However, combined use of melimine with ciprofloxacin at 0.5X MICs resulted in 91% inhibition of biofilm, and this inhibition was significantly higher ($p < 0.001$) than the 82% produced by melimine alone at 0.5X MICs (Figure 3A). Similarly, Mel4 and ciprofloxacin in combination at 0.5X MIC produced 83% inhibition of biofilm which was significantly higher ($p = 0.036$) than the 78% produced by Mel4 alone (Figure 3A). There was no significant difference in biofilm inhibition between melimine and ciprofloxacin, and Mel4 and ciprofloxacin combinations at 0.5X MIC ($p > 0.999$).

The biofilms produced by the ciprofloxacin-sensitive cells of ATCC 25923 were inhibited by $\geq 86\%$ by ciprofloxacin at ≥ 1X MIC ($p < 0.001$; Figure 3B). Melimine or Mel4 were active at 0.5X MICs and produced 82% and 78% biofilms inhibition compared to negative control, respectively ($p < 0.001$). The combinations of melimine or Mel4 with ciprofloxacin at 0.5X MIC produced reductions that were significantly higher (97%) than those used alone at 0.5X ($p < 0.001$). The combinations of either AMP with ciprofloxacin inhibited the same amount of biofilm at 0.5X MICs ($p > 0.999$; Figure 3B).

2.4. Disruption of Pre-Formed Biofilms by AMPs and Ciprofloxacin Alone or in Combination

In comparison to the effect of the AMPs or the combination of AMPs with ciprofloxacin on preventing the production of biofilms, all were less active in reducing pre-formed biofilms. For melimine or Mel4 at 0.5X to 2X MIC, pre-formed biofilms of either the ciprofloxacin-resistant or sensitive cells were 4–6 times more resistant than the biofilms formed in the presence of melimine.

The ability of AMPs and ciprofloxacin alone or in combination to disrupt pre-formed (24 h) biofilms of ciprofloxacin-resistant and sensitive isolates of *S. aureus* ATCC 25923 is presented in Figure 4. Ciprofloxacin did not reduce pre-formed biofilms of the ciprofloxacin-resistant isolate of *S. aureus* ATCC 25923 at any of the concentrations tested ($p > 0.999$; Figure 4A). Both AMPs reduced the amount of pre-formed biofilms in a concentration-dependent manner except at 0.5X MIC. Melimine produced 42%, 69% and 100% while Mel4 disrupted 38%, 64% and 97% at 1X, 2X and 4X MICs compared to negative control, respectively ($p < 0.001$; Figure 4A). Disruption of biofilm by melimine and Mel4 was similar at their corresponding MICs ($p > 0.999$). The combination of melimine and ciprofloxacin resulted in 69% biofilm disruption and the combination of Mel4 and ciprofloxacin resulted in 86% biofilm disruption at their corresponding 1X MIC compared to negative control ($p < 0.001$). The combined treatment of either AMP with ciprofloxacin at 1X MIC resulted in similar biofilm disruption ($p > 0.999$).

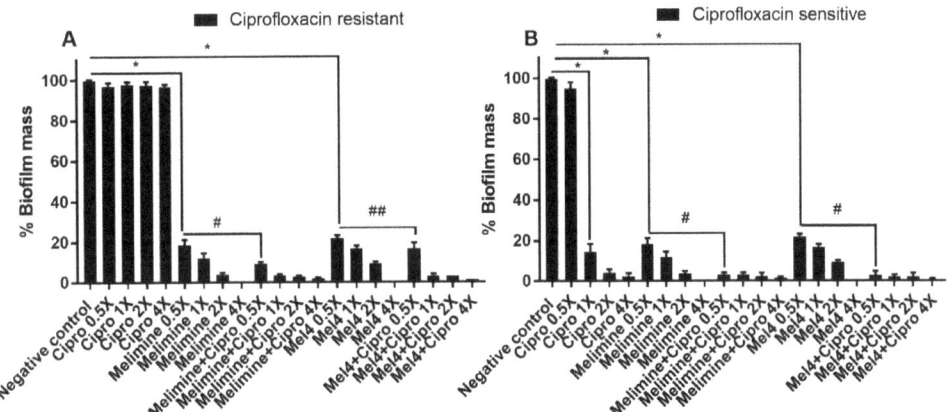

Figure 3. Inhibition of biofilm formation of *S. aureus* ATCC 25923. Biofilm formation of the ciprofloxacin-resistant (**A**) or sensitive (**B**) cells of *S. aureus* ATCC was inhibited by various concentrations of melimine, Mel4 and ciprofloxacin alone or in combination. The strain was made resistant to ciprofloxacin by sub-passage for 30 days at a sub-MIC concentration. * represent significant ($p < 0.001$) decreases compared to the negative control (bacteria grown in the absence of antibiotics). # indicates significant ($p < 0.001$) decrease for the combinations compared to melimine or Mel4 alone while ## indicates $p = 0.036$ compared to Mel4 alone. Means (±SD) of three independent repeats in triplicate. Negative control = bacteria grown in the absence of antimicrobials, Cipro = ciprofloxacin.

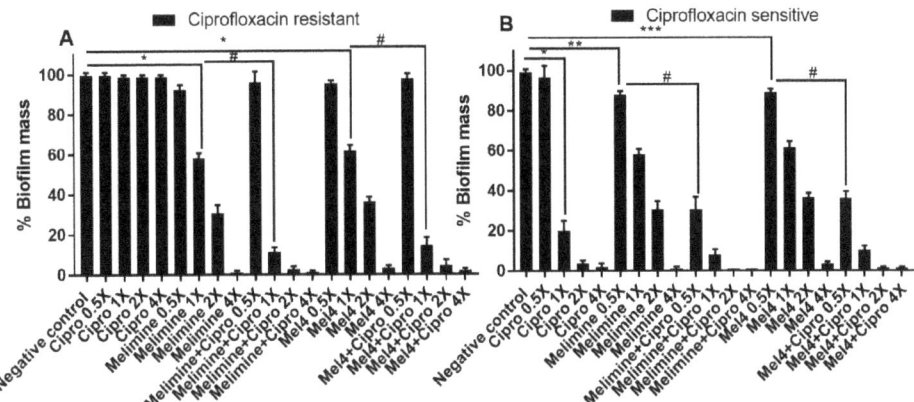

Figure 4. Disruption of pre-established biofilm of *S. aureus* ATCC. Biofilms of the ciprofloxacin-resistant (**A**) and sensitive (**B**) cells of *P. aeruginosa* ATCC 27853 were disrupted at various concentrations by melimine, Mel4 and ciprofloxacin alone or in combination. * represents significant ($p < 0.001$), ** indicates significant ($p = 0.005$), *** indicates significant ($p = 0.022$) decrease compared to the negative control (biofilm treated with buffer). # indicates significant ($p < 0.001$) decrease for the combinations compared to melimine or Mel4 alone. Error bars represent means (±SD) of three independent repeats in triplicate. Negative control = bacteria grown in the absence of antimicrobials. Cipro = ciprofloxacin.

Pre-formed biofilms of the ciprofloxacin-sensitive strain of S. *aureus* ATCC 25923 were susceptible to the action of ciprofloxacin at 1X MIC or higher concentrations. Ciprofloxacin disrupted pre-formed biofilms in a dose-dependent manner by producing 86%, 96% and 100% disruption of biofilms at 1X, 2X and 4X MICs, respectively, compared to control ($p < 0.001$; Figure 4B). Melimine disrupted 11% ($p = 0.005$) and Mel4 disrupted 10% ($p = 0.022$) of pre-formed biofilms compared to negative control at 0.5X MIC. At 1X MIC, melimine eradicated 41% of biofilm while Mel4 eradicated 37% of biofilm compared to buffer-treated negative controls (Figure 4B; $p < 0.001$). Interestingly, when AMPs were used in combination with ciprofloxacin, these combinations resulted in higher pre-formed biofilm disruption at concentrations lower than their MICs. The combination of melimine with ciprofloxacin at 0.5X MIC produced significantly higher (68%) biofilm disruption than when melimine (11%) was used alone at 0.5X (Figure 4B; $p < 0.001$). Similarly, the combination of Mel4 with ciprofloxacin at 0.5X MIC produced significantly higher (63%) biofilm disruption than when Mel4 (10%) was used alone at 0.5X (Figure 4B; $p < 0.001$). The combined treatment of either AMP with ciprofloxacin at 0.5X MIC resulted in similar biofilm disruption ($p > 0.999$). Similarly, at 1X MIC the combination of melimine with ciprofloxacin disrupted more highly (91%) than by melimine alone (41%) and Mel4 and ciprofloxacin disrupted more (89%) than by Mel4 alone (37%; $p < 0.001$). The combined antibiofilm effect of either peptide with ciprofloxacin was similar at 1X MIC ($p > 0.999$).

2.5. Visualization of Biofilms

Biofilms of the ciprofloxacin-resistant cells treated with buffer (HEPES) or ciprofloxacin alone had an overall dimension of 90 μm by 90 μm by 21 μm and the cells were mainly green, indicating that they were alive (Figure 5). Biofilms treated with melimine or Mel4 at 4X their MICs had less biofilm mass with dimensions of 43 μm by 43 μm by 6 μm and the cells were mainly stained red indicating many dead cells. No biofilms could be seen for the melimine and ciprofloxacin or Mel4 and ciprofloxacin combinations at 4X MICs (Figure 5).

2.6. Mechanistic Studies

2.6.1. Cell Membrane Depolarization

Melimine and Mel4 depolarized the cell membrane of S. *aureus* in biofilms in a concentration- and time-dependent manner (Figure 6A,B). Both peptides depolarized the cell membrane of biofilm cells within 1 h of incubation at 1X, 2X and 4X MICs. The fluorescence intensity produced as a result of the release of the DiSC3 (5) dye was higher at 4X than at 2X and 1X MIC for both melimine and Mel4 ($p \leq 0.004$). The rate of release of the dye increased up to 2 h and became constant thereafter for all concentrations. There was no difference in release of dye between melimine and Mel4 at their corresponding MICs ($p \geq 0.999$). Ciprofloxacin did not depolarize the cell membrane at any of the concentrations tested over the entire 6 h of the experiment. The combined membrane depolarizing effect of melimine or Mel4 with ciprofloxacin was almost exactly equivalent to the individual effects of melimine or Mel4 at their corresponding 1X, 2X, and 4X MICs ($p > 0.937$; Figure 6A,B). There was no difference between the combinations at 1X and 2X MICs ($p > 0.999$). However, at 4X MIC, the melimine and ciprofloxacin combination caused higher membrane depolarization than the Mel4 and ciprofloxacin combination after 2 h of incubation ($p = 0.005$). The positive control (DMSO 20%) gave maximum fluorescence at 2 h which became constant following this time point.

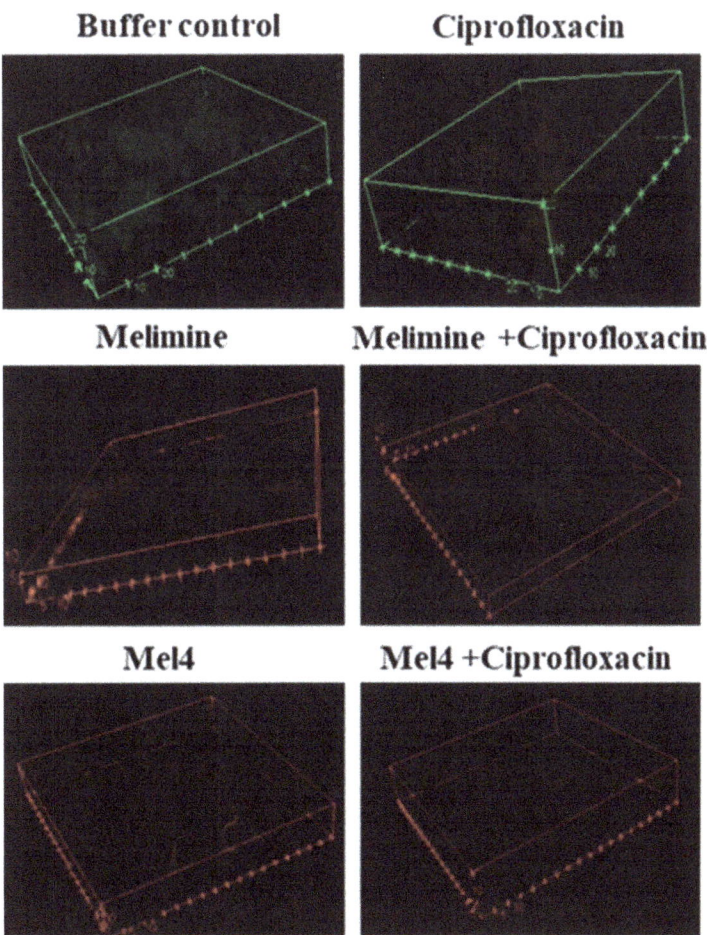

Figure 5. Representative confocal laser scanning microscopy images of biofilms of the ciprofloxacin resistant isolates of *S. aureus* ATCC 25923 after treatment with AMPs and ciprofloxacin alone or in combination. The antibiofilm effects were evaluated at 4X the MIC of all antimicrobials after incubation for 24 h. The biofilms of *S. aureus* were stained with SYTO-9 (excited at 488, green live cells) and propidium iodide (excited at 514 mm, red dead cells). The cells exposed to ciprofloxacin alone when excited at 514 nm had a reddish color indicating some of the cells had taken up the propidium iodide.

2.6.2. Release of Cellular Contents

Incubation of the AMPs with pre-formed biofilms of *S. aureus* ATCC 25923 released a substantial amount of ATP in a concentration-dependent manner (Figure 7). Melimine at 1X, 2X and 4X MIC induced leakage of 143 ± 15 nM, 167 ± 15 nM and 227 ± 21 nM ATP, respectively, compared to buffer-treated negative controls ($p < 0.001$). Mel4 at 1X, 2X and 4X MICs released 107 ± 25 nM, 142 ± 13 nM and 197 ± 21 nM extracellular ATP, respectively, compared to negative control ($p \leq 0.003$). The amount of ATP released by melimine and Mel4 at their corresponding MICs was similar ($p \geq 0.999$). The addition of ciprofloxacin alone to pre-formed biofilms did not result in the significant release of extracellular ATP at any of the concentrations tested ($p > 0.999$; Figure 7). However, the combination of melimine or Mel4 with ciprofloxacin resulted in the release of higher amounts of ATP than the AMPs alone. At 2X MIC, the melimine and ciprofloxacin combination released significantly higher amounts of ATP (233 ± 38 nM; p = 0.005) than released by melimine alone (167 ± 15 nM). There was similar effect on ATP leakage of the combination at 2X and 4X MICs. The combination of Mel4 and ciprofloxacin at 1X, 2X and 4X concentrations induced leakage of 152 ± 24 nM, 203 ± 32 nM and 267 ± 12 nM ATP, respectively (Figure 7). At 2X MIC, the combination of Mel4 and ciprofloxacin released significantly higher amounts of ATP (p = 0.002) than was released by Mel4 alone at 1X MIC. Both the melimine and ciprofloxacin or Mel4 and ciprofloxacin combination had similar effects at their corresponding MICs (p > 0.999).

The release of nucleic acids (260 nm absorbing material) after incubation for 4 h with the antimicrobials from pre-formed biofilms of *S. aureus* ATCC 25923 is shown in Figure 8A. Melimine released a significantly higher amount of DNA/RNA at 2X MIC (7 ± 1 times; p = 0.043) and 4X MIC (13 ± 1 times; p < 0.001) compared to control. Ciprofloxacin did not cause significant DNA/RNA leakage from the pre-formed biofilms at any concentration tested (p > 0.999; Figure 8A). The combination of melimine and ciprofloxacin released 10 ± 2 times (p = 0.047) more DNA/RNA compared to negative control at 2X MIC. Melimine and ciprofloxacin in combination released significantly higher (p = 0.022; Figure 8A) amounts of DNA/RNA than melimine alone at 2X MIC. The combination of Mel4 and ciprofloxacin did not release significant amounts of DNA/RNA at any concentration tested ($p \geq 0.480$). Melimine either alone or in combination with ciprofloxacin produced higher fluorescence at 2X and 4X MICs than other concentrations ($p \leq 0.034$; Figure 8B). Mel4 either alone or in combination with ciprofloxacin did not produce significant fluorescence at any concentration tested (p > 0.999; Figure 8B).

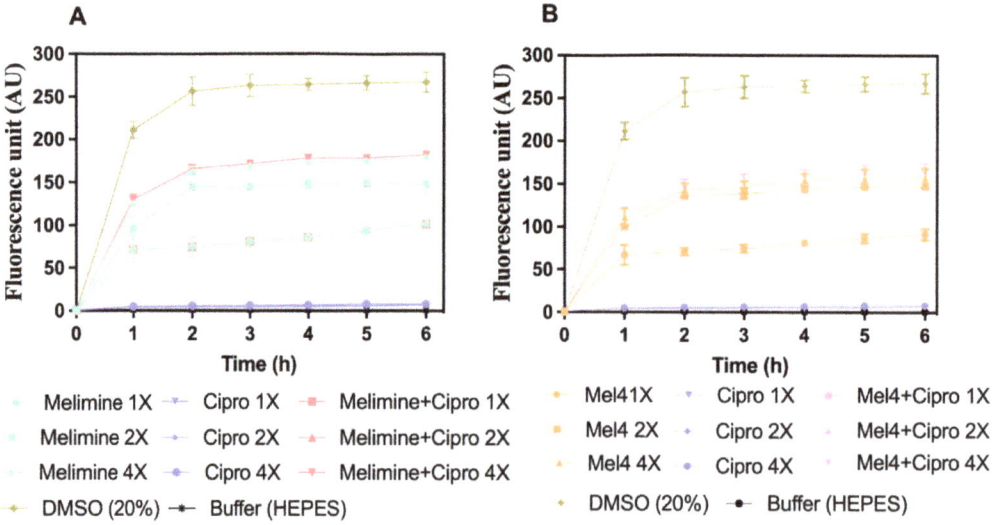

Figure 6. Cell membrane depolarization of pre-formed (24 h) biofilm cells. Cell membrane depolarization of *S. aureus* ATCC 25923 (became resistant to ciprofloxacin after 30 days of serial passages at sub-MIC) (**A**) by melimine and ciprofloxacin alone or in combination, and (**B**) by Mel4 and ciprofloxacin alone or in combination against pre-formed (24 h) biofilms. Error bars are means (±SD) of three independent repeats in triplicate. Cipro = ciprofloxacin, DMSO = dimethyl sulfoxide.

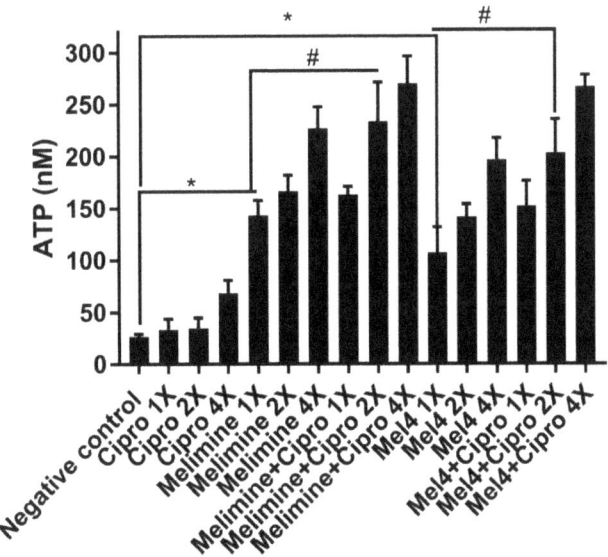

Figure 7. Leakage of ATP from pre-formed biofilm cells of *S. aureus* ATCC 25923. Leakage of ATP following treatment for 3 h with either of the two peptides and ciprofloxacin alone or in combination. The strain was made resistant to ciprofloxacin by passage for 30 days at a sub-MIC. * represents significant ($p < 0.001$) increases in the amount of extracellular at inhibitory concentrations of peptides ATP compared to the negative control. # represents significant ($p < 0.001$) increase in the release of ATP of the combination of melimine or Mel4 with ciprofloxacin compared to melimine or Mel4 alone.

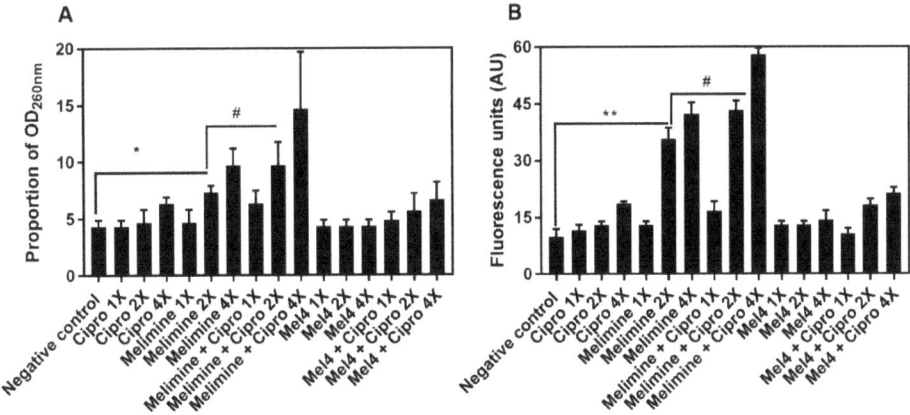

Figure 8. Increase in OD_{260nm} after release of DNA/RNA (**A**) and increase in fluorescence after interaction of Sytox green with released DNA/RNA (**B**) from pre-formed biofilm cells of *S. aureus* ATCC 25923. Leakage of nucleic acid from pre-formed (24 h) biofilms of *S. aureus* ATCC 25923 following treatments for 3 h with either of the two peptides and ciprofloxacin alone or in combination. The strain was made resistant to ciprofloxacin by passage of 30 days at a sub-MIC concentration. * represents significance ($p = 0.043$) and ** indicates ($p \leq 0.034$) release of nucleic acid compared to the negative control. # represents significant ($p = 0.022$) increase in the release of nucleic acid by the combination of melimine and ciprofloxacin compared to melimine alone.

3. Discussion

Exposure of bacteria to sub-inhibitory concentrations of antimicrobials can result in generation of resistant mutants [26,27]. The current study demonstrated that the AMPs melimine and Mel4 at sub-MICs did not induce resistance in *S. aureus* ATCC 25923. We and others [28–31] have tested several broad-spectrum antibiotics such as gentamicin (data not shown in the current study) and ciprofloxacin to determine whether strains such as *S. aureus* ATCC 6538, ATCC 25923, 31 and 38 can develop resistance to gentamycin and ciprofloxacin. Resistance to gentamicin or ciprofloxacin was not induced in any strain except *S. aureus* ATCC 25923 which developed resistance against ciprofloxacin. Therefore, ciprofloxacin was selected to determine its activity alone or in combination with antimicrobial peptides against this strain. Biofilms of the resistance cells of *S. aureus* ATCC 25923 could be reduced by treatment with combinations of melimine or Mel4 with ciprofloxacin whilst the biofilm was forming or once it had developed.

S. aureus ATCC 25923 developed resistances to ciprofloxacin similar to *P. aeruginosa* ATCC 27853 [13], in a step-wise manner to full resistance (>120X MIC) after 25 days of passage. Resistance to ciprofloxacin in *S. aureus* can occur due to mutations in *grlA/grlB* and *gyrA/gyrB* genes, which encode the subunits of topoisomerase IV and DNA gyrase, respectively [32,33], or over expression of the membrane-associated protein NorA efflux pump which leads to increased transport of ciprofloxacin out of the bacterial cell [34]. Changes in these genes may occur randomly during exposure to ciprofloxacin and this may be why the resistance occurs sporadically during exposure to the antibiotic. In contrast to *S. aureus* ATCC 25923, all other *S. aureus* strains (31, 38 and ATCC 6538) did not mutate and develop resistance against ciprofloxacin. None of the *S. aureus* strains was able to develop resistance against melimine and Mel4. The inability of *S. aureus* to develop resistance against melimine and Mel4 may be due to the rapid killing kinetics of these peptides and action on cell membranes [23]. Bacteria appear to rarely gain resistance to AMPs that target bacterial membranes [23,35]. However, like other Gram-positive bacteria, *S. aureus* can develop resistance to AMPs by reducing the negative charge on teichoic acid and production of proteases that fragment AMPs [36,37], but these mechanisms appear not to have been activated during growth in sub-MICs of melimine or Mel4.

Another mechanism whereby bacteria can protect themselves from the action of antimicrobials is formation of biofilms [38]. Melimine and Mel4 prevented biofilm formation of *S. aureus* at a concentration lower than their MICs. A similar effect has been shown with the cathelicidin-derived peptide NA-CATH:ATRA1-ATRA1 against *S. aureus* biofilm [39]. The AMPs esculentin-3, Tet-213 and 1010 peptides prevent biofilm formation [40,41] by stimulating twitching motility, influencing quorum sensing or degrading signaling molecules such as ppGpp which lead to changes in the expression of genes related to biofilm assembly [42–44].

Melimine and Mel4 killed biofilm cells and dispersed pre-formed biofilms. Similarly, AMPs such as LL37, DL-K6L9, Seg5L, Seg5D, Seg6L, and Seg6D killed the biofilm cells and reduced the biofilm mass by dispersing the biofilm matrix [45,46]. Both our AMPs followed a similar mechanism, as treating biofilms of ciprofloxacin-resistant cells with either AMP resulted in a high proportion of PI positive (stained red = dead cells) with a reduced biofilm mass compared to buffer-treated negative controls. Disruption of pre-formed biofilm by these two AMPs was similar to disruption of pre-formed biofilm of *P. aeruginosa* [13]. Like the case with *P. aeruginosa*, the anti-biofilm effects of melimine and Mel4 against *S. aureus* were similar to their mode of action on *S. aureus* cells in suspension [23]; this is depolarization of membranes and release of intracellular contents.

However, the speed of the effects of melimine and Mel4 was decreased compared to their effects on planktonic cells [13], which may be due to the complex structure of *S. aureus* biofilms hindering the antimicrobial action of AMPs. Membrane depolarization of biofilm cells caused by melimine and Mel4 was slower and happened after one hour compared to only 30 s against planktonic bacteria [23]. Similarly, membrane depolarization of *S. aureus* cells in biofilms occurred after 1 h with the AMPs nisin A and lacticin Q [47]. The time required to depolarize the membrane of *S. aureus* biofilm cells was similar to *P. aeruginosa* biofilm cells [13]. Slower membrane depolarization of biofilm cells compared to planktonic bacteria might be due to higher viscosity of biofilm which can affect the penetration of AMPs in biofilm [47–49]. Moreover, negatively charged polymers of biofilms may interact with the positively charged AMPs and limit penetration and diffusion of AMPs in biofilm matrix.

Both AMPs killed the biofilm cells by damaging the membranes followed by leakage of cellular ATP. Leakage of ATP from biofilm cells was slower and occurred after 3 h compared to after 2 min from planktonic bacteria [23]. As discussed above, this change in timing of events may be due to the charge of biofilm polymers or viscosity within biofilms. Higher concentrations of AMPs above their MICs may disrupt the membrane of biofilm cells to a greater extent and start to release larger molecules [48,50–52]. Melimine released DNA/RNA from biofilm cells at 4X MIC. On the other hand, Mel4 alone or in combination with ciprofloxacin did not result in release of DNA/RNA even at 4X its MIC. The mechanism of action of Mel4 against biofilm cells seems to be similar to planktonic cells which are independent of the release of DNA/RNA [23].

The combination of AMPs and ciprofloxacin inhibited greater biofilm formation at 0.5X than alone, suggesting that both the peptides may have additive or synergistic effects against *S. aureus*. The AMPs indolicidin, cecropin (1–7) and nisin in combination with ciprofloxacin inhibited the *S. aureus* biofilm at concentrations lower than their MICs [38]. The fractional inhibitory concentrations of these AMPs with ciprofloxacin were above synergistic levels, showing additive effects instead, against planktonic *S. aureus* [24]. The combination of AMPs with ciprofloxacin resulted in more biofilm disruption at 1X MIC than alone. These results coincide with the previous study which reported that the AMPs indolicidin, cecropin (1–7)–melittin A (2–9) and nisin in combination with teicoplanin or ciprofloxacin disrupted the biofilm of methicillin-resistant *S. aureus* at 1X MIC [53]. Smaller differences in biofilms inhibition/disruption may be due to sensitivity of the strain towards antibiotics, maturation of biofilms and concentration of antimicrobials used. Several peptides in combination with antibiotics have been tested against biofilms formed for 2 h to 4 h, at concentrations 2–4 times lower than their MICs. Table 2 compares these combinations with melimine or Mel4 with ciprofloxacin tested at their 0.5X MICs against biofilms formed for 24 h in the present study. The slightly higher effects of the combination of Citropin1.1 + Minocycline [54] or LL37 + Teicoplanin [20] may be due to the fact the biofilms were only produced for 4 h, whereas the current study used biofilms formed over 24 h and these longer times might produce more robust biofilms. The effect of both the peptides with ciprofloxacin against *S. aureus* biofilm is summarized in Figure 9. The ability of the AMP-ciprofloxacin combinations to disrupt greater amounts of pre-formed biofilms might be related to AMPs' facilitating higher intracellular uptake of ciprofloxacin [55]. The AMPs WR12, SAAP-148, SAAP-276 and TC84 allowed greater cellular uptake of ciprofloxacin and teicoplanin by permeabilizing the cell membrane of *S. aureus* in biofilms [20,55]. Another possible mechanism of AMP-antibiotic combinations is disrupting the biofilm matrix to allow AMPs to target the bacterial cells in the biofilm and cause dispersion of cells in the biofilm [56].

Table 2. Effect of antimicrobial peptides and antibiotics at 0.5X MIC in combination against *S. aureus* biofilm.

Antimicrobial Agents	Biofilm Inhibition (%)	Biofilm Eradication (%)
Melimine + Ciprofloxacin	91%	69%
Mel4 + Ciprofloxacin	83%	86%
Citropin1.1 + Minocycline [54]	>99%	ND
Indolicidin + Daptomycin [53]	44%	ND
Nisin + Ciprofloxacin [53]	50%	ND
LL37 + Teicoplanin [20]	ND	>99%
Temporin A + Gentamycin [57]	ND	90%
Indolicidin + Ciprofloxacin [38]	ND	47%

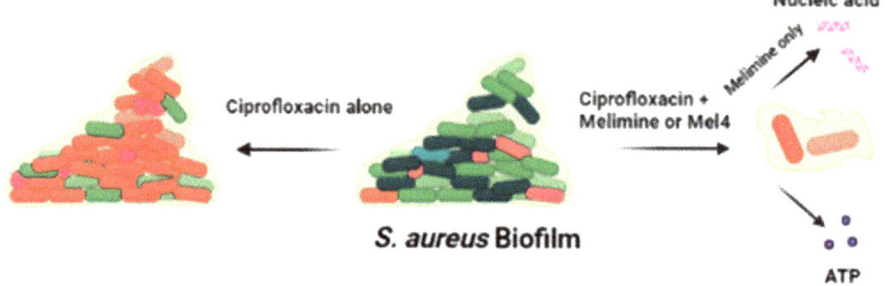

Figure 9. Effect of ciprofloxacin and peptides on the pre-formed biofilm of *S. aureus*. Ciprofloxacin alone did not disrupt the biofilm while when in combination with melimine or Mel4 it destroys the biofilm matrix following release of DNA/RNA (with melimine only) and ATP from biofilm cells.

4. Materials and Methods

4.1. Synthesis of Peptides and Bacteria

Melimine and Mel4 were synthesized by conventional solid-phase peptide protocol [58,59] and were procured from the Auspep Peptide Company (Tullamarine, Victoria, Australia). The purity of the peptides was ≥90%. Ciprofloxacin was purchased from Sigma-Aldrich (St Louis, MO, USA). Ciprofloxacin stock solution (5120 µg/mL) in milli Q water was prepared and stored at −30 °C. Bacterial strains such as *S. aureus* 31 (mecA positive) and *S. aureus* 38 (mecA negative; both microbial keratitis isolates) [60] and two reference strains *S. aureus* ATCC 6538 (mecA negative; a human lesion isolate) and *S. aureus* ATCC 25923 were used in the current study.

4.2. Minimal Inhibitory Concentration and Minimal Bactericidal Concentration

The minimum inhibitory and minimum bactericidal concentrations of ciprofloxacin were determined using a standard broth microdilution method of the Clinical Laboratory and Standard Institute (CLSI) and a modified version of the CLSI broth microdilution method was used to determine the MIC of antimicrobial peptides [61]. The MIC was set as the lowest concentration that reduced bacterial growth by ≥90% while the MBC was set as the lowest concentration that reduced bacterial growth by >99.99% following enumeration of live bacteria by plate counts compared to bacteria grown in the absence of any antimicrobial.

4.3. Growth Curve and Resistance Development at Sub-MIC of Antimicrobials

An aliquot (100 µL) of an overnight culture (1×10^6 CFU/mL) of bacteria was added to an equal volume of each antimicrobial to achieve a sub-MIC (0.5X MIC) in MHB and was incubated at 37 °C with shaking at 120 rpm for 24 h. The turbidity of the bacterial suspensions was determined at OD_{660nm} over time for 24 h. Bacteria grown in wells without antimicrobials served as positive controls for maximum bacterial growth. Serial passages of *S. aureus* ATCC 25923 were performed in the presence of each antimicrobial at 0.5X MIC. After incubation for 18–24 h, cells were repassaged into fresh media containing sub-MICs of the antimicrobials. After every passage, the MIC for each antimicrobial was determined, and a new sub-MIC was adjusted if any increase in MIC was observed. This repassaging lasted for 30 consecutive days. *S. aureus* 31, *S. aureus* 38, *S. aureus* ATCC 6538 and *S. aureus* ATCC 25923 strains were exposed to AMPs and ciprofloxacin at sub-MIC (one-fold below the MIC) for their ability to develop resistance against these antimicrobials. Of all the tested strains, only *S. aureus* ATCC 25923 was able to develop resistance to ciprofloxacin using this method. This strain has been shown to be able to develop resistance to ciprofloxacin previously [28].

4.4. Inhibition of Biofilm Formation by AMPs and Ciprofloxacin Alone or in Combination

Inhibition of biofilm formation by AMPs alone or in combination with ciprofloxacin was determined using *S. aureus* 25923 that had been passaged for one day (sensitive cells) or thirty days (resistant cells). First, 100 µL of *S. aureus* (1×10^6 CFU/mL) was dispensed into round-bottom 96-well microtiter plates containing serial dilutions (0.5X to 4X MIC) of melimine, Mel4 or ciprofloxacin. Then plates were incubated at 37 °C with shaking at 120 rpm for 24 h. The combined effect of melimine or Mel4 with ciprofloxacin was determined after adding equal volumes of each at their corresponding MICs. Wells containing bacteria and MHB and treated with buffer served as negative controls. Following incubation, the media were removed, and wells were then carefully washed two times with HEPES buffer to remove non-adherent cells. Subsequently, biofilms were fixed with 200 µL of 99% v/v methanol for 15 min and then plates were air dried. Finally, biofilms were stained with 200 µL of 1% w/v crystal violet dissolved in water for 5 min. Unbound crystal violet was rinsed off with tap water and plates were inverted to air dry. The crystal violet absorbed in biofilms was solubilized in 200 µL glacial acetic acid (33%, v/v), the released dye was moved to new well and the amount of dye released was determined spectroscopically at OD_{600nm}. The degree of biofilm inhibition was determined as a percentage of the biofilm produced by the negative controls (bacteria with no antimicrobials) using the following formulae [62].

$$\% \text{ biofilm of single or combined antimicrobial} = \frac{(OD_{600nm} \text{ of negative control}) - (OD_{600nm} \text{ of individual or (combined) antimicrobials})}{(OD_{600nm} \text{ of negative control})} \times 100 \quad (1)$$

4.5. Disruption of Pre-Formed Biofilms by AMPs and Ciprofloxacin Alone or in Combination

Biofilms were formed by adding 100 µL of *S. aureus* ATCC 25923 (1×10^6 CFU/mL) ciprofloxacin-sensitive or resistant cells into round-bottom 96-well microtiter plates containing 100 µL of MHB. Plates were incubated at 37 °C in static condition. After incubation, biofilms were treated with serially diluted peptides or ciprofloxacin or their combination at their corresponding MICs and the plates were incubated for a further 24 h at 37 °C in static condition. Wells containing bacteria and MHB and treated with buffer served as negative controls. Following incubation, the media were removed, and wells were then carefully washed two times with HEPES buffer to remove non-adherent cells and the amount of biofilm was determined as outlined in the previous experiment.

The ability of each antimicrobial to disrupt pre-formed biofilms formed by resistant (30-day ciprofloxacin-passaged) *S. aureus* ATCC 25923 was visualized with confocal laser scanning microscopy (FV 1200, Olympus, Tokyo, Japan). A 24 h pre-formed biofilm on sterile round glass coverslips in polystyrene plates was treated with 200 µL of 4X-MIC of melimine, Mel4 or ciprofloxacin alone or in combination at 37 °C for 24 h. Thereafter, biofilms were stained with Live/Dead BacLight bacterial viability kit (Invitrogen, Eugene, OR, USA) and examined with confocal microscopy. The resulting data were processed using the Image J software version 8 (Bethesda, MD, USA).

4.6. Mechanistic Studies

As both AMPs had similar antibiofilm effects against either 1-day or 30-day ciprofloxacin-passaged strains of *S. aureus* ATCC 25923, the 30-day ciprofloxacin-passaged cells were selected to evaluate the mechanism of action of both the AMPs and ciprofloxacin towards bacterial cells in biofilms.

4.7. Effect on Cell Membranes

The depolarizing effect on the cell membranes of biofilm-embedded cells was determined as described previously [48].

Briefly, 24 h formed biofilms were washed with 5 mM HEPES (pH 7.2) containing 20 mM glucose and 100 mM KCl at pH 7.2. Then, biofilm cells were loaded with the membrane potential sensitive dye DiSC3 (5) (4 µM; Sigma Aldrich, St Louis, MO, USA)) in HEPES for 1 h in dark. Release of DiSC3 (5) following addition of serially diluted melimine, Mel4 or ciprofloxacin alone or in combination at 1X, 2X and 4X their respective MICs was recorded at regular intervals up to 6 h. DMSO (20%; Merck, Billerica, MA, USA) was used as a positive control to achieve maximum membrane depolarization.

4.8. Release of Cellular Contents

The biofilm cells were incubated with serially diluted melimine, Mel4 or ciprofloxacin alone or in combination at 1X, 2X and 4X their corresponding MICs. The supernatants were removed after 3 h and filtered through 0.22 µm pore membranes (Merck, Tullagreen, Ireland). Subsequently, the amount of extracellular of ATP was measured using a bioluminescence kit (Invitrogen, Eugene, OR, USA) according to manufacturer's instructions. Buffer (HEPES)-treated samples were used as negative controls [47].

Similarly, supernatant was also analyzed for release of nucleic acids (DNA/RNA) [26]. The supernatants were centrifuged at $1300 \times g$ for 10 min and then filtered through 0.22 µm pore membranes (Merck). The OD_{260nm} of the filtrate was measured, and the results were expressed relative to the initial OD_{260nm} of biofilms taken at 0 min. Furthermore, the presence of nucleic acids in the supernatants was also confirmed with Sytox green (5 µM Invitrogen, Eugene, OR, USA) as final concentration. An increase in fluorescence due to the interaction of Sytox green with nucleic acid was measured spectrophotometrically at an excitation wavelength of 480 nm and an emission wavelength of 523 nm.

4.9. Statistical Analysis

All experiments were performed in three independent assays. One-way analysis of variance (ANOVA) with Bonferroni's corrections for multiple comparisons was used to compare differences between control and antimicrobial-treated cells. The data of cell membrane depolarization were analyzed using two-way ANOVA with Tukey's test. A probability value of $p < 0.05$ was considered statistically significant.

5. Conclusions

In conclusion, *S. aureus* in suspension could not become resistant to melimine or Mel4 following repeated exposure in sub-inhibitory concentrations of these AMPs. Whilst both AMPs inhibited biofilm formation, once *S. aureus* had produced a biofilm, the cells became more resistant to melimine or Mel4, although they could still act against the biofilms at 4X their MICs. Moreover, the combination of the AMPs and ciprofloxacin produced greater effects, possibly as a result of the AMPs damaging the cell membrane of biofilm cells which resulted in increased or facilitated uptake of ciprofloxacin. Future research should be conducted, using, for example, fluorescently labelled ciprofloxacin to examine whether the combination results in greater uptake of ciprofloxacin.

Author Contributions: M.Y. designed the study, performed the experiments, analyzed the data and wrote the manuscript; D.D. supervised and helped M.Y. in analyzing data and edited the article. M.D.P.W. planned the project, developed the theoretical framework and edited the article. All authors have read and agreed to the published version of the manuscript.

Funding: This work was supported by the Australian Research Council (ARC) under Grant DP160101664 and the National Health and Medical Research Council under grant APP1183597. First author acknowledges the UNSW and HEC Pakistan for provision of tuition fee scholarship and living allowance respectively.

Acknowledgments: The authors acknowledge the facilities and the technical assistance provided by Michael J Carnell Biomedical Imaging Facility (BMIF) at the University of New South Wales, Australia for help with the confocal microscopy.

Conflicts of Interest: This work is original, has not been published and is not being considered for publication elsewhere. There are no conflicts of interest for any of the authors that could have influenced the results of this work. The funders had no role in the design of the study; in the collection, analyses, or interpretation of data; in the writing of the manuscript, or in the decision to publish the results.

References

1. Manandhar, S.; Singh, A.; Varma, A.; Pandey, S.; Shrivastava, N. Biofilm producing clinical *Staphylococcus aureus* isolates augmented prevalence of antibiotic resistant cases in Tertiary Care Hospitals of Nepal. *Front. Microbiol.* **2018**, *9*, 2749. [CrossRef]
2. Idrees, M.; Sawant, S.; Karodia, N.; Rahman, A. *Staphylococcus aureus* biofilm: Morphology, genetics, pathogenesis and treatment strategies. *Int. J. Environ. Res. Public Health* **2021**, *18*, 7602. [CrossRef]
3. Health, U.D.o.; Services, H. *Antibiotic Resistance Threats in the United States*; CDC: Atlanta, GA, USA, 2013.
4. Neopane, P.; Nepal, H.P.; Shrestha, R.; Uehara, O.; Abiko, Y. In vitro biofilm formation by *Staphylococcus aureus* isolated from wounds of hospital-admitted patients and their association with antimicrobial resistance. *Int. J. Gen. Med.* **2018**, *11*, 25–32. [CrossRef] [PubMed]
5. Mohammad, H.; Thangamani, S.; Seleem, M.N. Antimicrobial peptides and peptidomimetics-potent therapeutic allies for staphylococcal infections. *Curr. Pharma. Des.* **2015**, *21*, 2073–2088. [CrossRef] [PubMed]
6. Stryjewski, M.E.; Chambers, H.F. Skin and soft-tissue infections caused by community-acquired methicillin-resistant *Staphylococcus aureus*. *Clin. Infect. Dis.* **2008**, *46*, S368–S377. [CrossRef] [PubMed]
7. Jaśkiewicz, M.; Janczura, A.; Nowicka, J.; Kamysz, W. Methods used for the eradication of staphylococcal biofilms. *Antibiotics* **2019**, *8*, 174. [CrossRef]
8. Høiby, N.; Bjarnsholt, T.; Moser, C.; Bassi, G.L.; Coenye, T.; Donelli, G.; Hall-Stoodley, L.; Holá, V.; Imbert, C.; Kirketerp-Møller, K.; et al. ESCMID guideline for the diagnosis and treatment of biofilm infections 2014. *Clin. Microbiol. Infect.* **2015**, *21*, S1–S25. [CrossRef]
9. Maya, I.D.; Carlton, D.; Estrada, E.; Allon, M. Treatment of dialysis catheter–related *Staphylococcus aureus* bacteremia with an antibiotic lock: A quality improvement report. *Am. J. Kidney Dis.* **2007**, *50*, 289–295. [CrossRef]
10. Liu, J.; Madec, J.-Y.; Bousquet-Mélou, A.; Haenni, M.; Ferran, A.A. Destruction of *Staphylococcus aureus* biofilms by combining an antibiotic with subtilisin A or calcium gluconate. *Sci. Rep.* **2021**, *11*, 6225. [CrossRef]
11. Batoni, G.; Maisetta, G.; Esin, S. Antimicrobial peptides and their interaction with biofilms of medically relevant bacteria. *Biochim. Biophys. Acta (BBA)-Biomembr.* **2016**, *1858*, 1044–1060. [CrossRef]
12. Grassi, L.; Maisetta, G.; Esin, S.; Batoni, G. Combination strategies to enhance the efficacy of antimicrobial peptides against bacterial biofilms. *Front. Microbiol.* **2017**, *8*, 2409. [CrossRef]
13. Yasir, M.; Dutta, D.; Willcox, M.D.P. Activity of antimicrobial peptides and ciprofloxacin against *Pseudomonas aeruginosa* biofilms. *Molecules* **2020**, *25*, 3843. [CrossRef] [PubMed]
14. Yasir, M.; Willcox, M.D.P.; Dutta, D. Action of antimicrobial peptides against bacterial biofilms. *Materials* **2018**, *11*, 2468. [CrossRef] [PubMed]
15. Mishra, N.M.; Briers, Y.; Lamberigts, C.; Steenackers, H.; Robijns, S.; Landuyt, B.; Vanderleyden, J.; Schoofs, L.; Lavigne, R.; Luyten, W.; et al. Evaluation of the antibacterial and antibiofilm activities of novel CRAMP–vancomycin conjugates with diverse linkers. *Org. Biomol. Chem.* **2015**, *13*, 7477–7486. [CrossRef] [PubMed]
16. Rudilla, H.; Fusté, E.; Cajal, Y.; Rabanal, F.; Vinuesa, T.; Viñas, M. Synergistic antipseudomonal effects of synthetic peptide AMP38 and carbapenems. *Molecules* **2016**, *21*, 1223. [CrossRef]

17. Ribeiro, S.M.; de la Fuente-Núñez, C.; Baquir, B.; Faria-Junior, C.; Franco, O.L.; Hancock, R.E.W. Antibiofilm peptides increase the susceptibility of carbapenemase-producing *Klebsiella pneumoniae* clinical isolates to β-lactam antibiotics. *Antimicrob. Agents Chemother.* **2015**, *59*, 3906–3912. [CrossRef] [PubMed]
18. Li, X.; Sun, L.; Zhang, P.; Wang, Y. Novel approaches to combat medical device-associated biofilms. *Coatings* **2021**, *11*, 294. [CrossRef]
19. Zharkova, M.S.; Orlov, D.S.; Golubeva, O.Y.; Chakchir, O.B.; Eliseev, I.E.; Grinchuk, T.M.; Shamova, O.V. Application of antimicrobial peptides of the innate immune system in combination with conventional antibiotics—A novel way to combat antibiotic resistance? *Front. Cell. Infect. Microbiol.* **2019**, *9*, 128. [CrossRef]
20. Koppen, B.C.; Mulder, P.P.G.; de Boer, L.; Riool, M.; Drijfhout, J.W.; Zaat, S.A.J. Synergistic microbicidal effect of cationic antimicrobial peptides and teicoplanin against planktonic and biofilm-encased *Staphylococcus aureus*. *Int. J. Antimicrob. Agents* **2019**, *53*, 143–151. [CrossRef]
21. Willcox, M.; Hume, E.; Aliwarga, Y.; Kumar, N.; Cole, N. A novel cationic-peptide coating for the prevention of microbial colonization on contact lenses. *J. Appl. Microbiol.* **2008**, *105*, 1817–1825. [CrossRef]
22. Dutta, D.; Cole, N.; Kumar, N.; Willcox, M.D.P. Broad spectrum antimicrobial activity of melimine covalently bound to contact lenses. *Investig. Ophthalmol. Vis. Sci.* **2013**, *54*, 175–182. [CrossRef]
23. Yasir, M.; Dutta, D.; Willcox, M.D.P. Mode of action of the antimicrobial peptide Mel4 is independent of *Staphylococcus aureus* cell membrane permeability. *PLoS ONE* **2019**, *14*, e0215703. [CrossRef]
24. Kampshoff, F.; Willcox, M.D.P.; Dutta, D. A pilot study of the synergy between two antimicrobial peptides and two common antibiotics. *Antibiotics* **2019**, *8*, 60. [CrossRef] [PubMed]
25. Stefan, C.P.; Koehler, J.W.; Minogue, T.D. Targeted next-generation sequencing for the detection of ciprofloxacin resistance markers using molecular inversion probes. *Sci. Rep.* **2016**, *6*, 25904. [CrossRef] [PubMed]
26. Andersson, D.I.; Hughes, D. Microbiological effects of sublethal levels of antibiotics. *Nat. Rev. Microbiol.* **2014**, *12*, 465–478. [CrossRef] [PubMed]
27. Vasilchenko, A.S.; Rogozhin, E.A. Sub-inhibitory effects of antimicrobial peptides. *Front. Microbiol.* **2019**, *10*, 1160. [CrossRef]
28. Pollard, J.E.; Snarr, J.; Chaudhary, V.; Jennings, J.D.; Shaw, H.; Christiansen, B.; Wright, J.; Jia, W.; Bishop, R.E.; Savage, P.B. In vitro evaluation of the potential for resistance development to ceragenin CSA-13. *J. Antimicrob. Chemother.* **2012**, *67*, 2665–2672. [CrossRef]
29. Campion, J.J.; McNamara, P.J.; Evans, M.E. Evolution of ciprofloxacin-resistant *Staphylococcus aureus* in in vitro pharmacokinetic environments. *Antimicrob. Agents Chemother.* **2004**, *48*, 4733–4744. [CrossRef]
30. Tuchscherr, L.; Kreis, C.A.; Hoerr, V.; Flint, L.; Hachmeister, M.; Geraci, J.; Bremer-Streck, S.; Kiehntopf, M.; Medina, E.; Kribus, M.; et al. *Staphylococcus aureus* develops increased resistance to antibiotics by forming dynamic small colony variants during chronic osteomyelitis. *J. Antimicrob. Chemother.* **2016**, *71*, 438–448. [CrossRef]
31. Bidossi, A.; Bottagisio, M.; Logoluso, N.; De Vecchi, E. In vitro evaluation of gentamicin or vancomycin containing bone graft substitute in the prevention of orthopedic implant-related infections. *Int. J. Mol. Sci.* **2020**, *21*, 9250. [CrossRef]
32. Fournier, B.; Hooper, D.C. Mutations in topoisomerase IV and DNA gyrase of *Staphylococcus aureus*: Novel pleiotropic effects on quinolone and coumarin activity. *Antimicrob. Agents Chemother.* **1998**, *42*, 121–128. [CrossRef] [PubMed]
33. Takahashi, H.; Kikuchi, T.; Shoji, S.; Fujimura, S.; Lutfor, A.B.; Tokue, Y.; Nukiwa, T.; Watanabe, A. Characterization of gyrA, gyrB, grlA and grlB mutations in fluoroquinolone-resistant clinical isolates of *Staphylococcus aureus*. *J. Antimicrob. Chemother.* **1998**, *41*, 49–57. [CrossRef] [PubMed]
34. Yoshida, H.; Bogaki, M.; Nakamura, S.; Ubukata, K.; Konno, M. Nucleotide sequence and characterization of the Staphylococcus aureus norA gene, which confers resistance to quinolones. *J. Bacteriol.* **1990**, *172*, 6942–6949. [CrossRef] [PubMed]
35. Bechinger, B.; Gorr, S.-U. Antimicrobial peptides: Mechanisms of action and resistance. *J. Dent. Res.* **2017**, *96*, 254–260. [CrossRef] [PubMed]
36. Peschel, A.; Otto, M.; Jack, R.W.; Kalbacher, H.; Jung, G.; Götz, F. Inactivation of the dlt operon in *Staphylococcus aureus* confers sensitivity to defensins, protegrins, and other antimicrobial peptides. *J. Biol. Chem.* **1999**, *274*, 8405–8410. [CrossRef]
37. Sieprawska-Lupa, M.; Mydel, P.; Krawczyk, K.; Wójcik, K.; Puklo, M.; Lupa, B.; Suder, P.; Silberring, J.; Reed, M.; Pohl, J.; et al. Degradation of human antimicrobial peptide LL-37 by *Staphylococcus aureus*-derived proteinases. *Antimicrob. Agents Chemother.* **2004**, *48*, 4673–4679. [CrossRef]
38. Dosler, S.; Mataraci, E. In vitro pharmacokinetics of antimicrobial cationic peptides alone and in combination with antibiotics against methicillin resistant *Staphylococcus aureus* biofilms. *Peptides* **2013**, *49*, 53–58. [CrossRef]
39. Dean, S.N.; Bishop, B.M.; van Hoek, M.L. Natural and synthetic cathelicidin peptides with anti-microbial and anti-biofilm activity against *Staphylococcus aureus*. *BMC Microbiol.* **2011**, *11*, 114. [CrossRef]
40. Gao, G.; Lange, D.; Hilpert, K.; Kindrachuk, J.; Zou, Y.; Cheng, J.T.; Kazemzadeh-Narbat, M.; Yu, K.; Wang, R.; Straus, S.K.; et al. The biocompatibility and biofilm resistance of implant coatings based on hydrophilic polymer brushes conjugated with antimicrobial peptides. *Biomaterials* **2011**, *32*, 3899–3909. [CrossRef]
41. Luca, V.; Stringaro, A.; Colone, M.; Pini, A.; Mangoni, M.L. Esculentin(1-21), an amphibian skin membrane-active peptide with potent activity on both planktonic and biofilm cells of the bacterial pathogen *Pseudomonas aeruginosa*. *Cell. Mol. Life Sci.* **2013**, *70*, 2773–2786. [CrossRef]
42. Mah, T.-F.C.; O'toole, G.A. Mechanisms of biofilm resistance to antimicrobial agents. *Trends Microbiol.* **2001**, *9*, 34–39. [CrossRef]

43. De la Fuente-Núñez, C.; Reffuveille, F.; Haney, E.F.; Straus, S.K.; Hancock, R.E. Broad-spectrum anti-biofilm peptide that targets a cellular stress response. *PLoS Pathog.* **2014**, *10*, e1004152. [CrossRef]
44. Dostert, M.; Belanger, C.R.; Hancock, R.E.W. Design and assessment of anti-biofilm peptides: Steps toward clinical application. *J. Innate Immun.* **2019**, *11*, 193–204. [CrossRef] [PubMed]
45. Kang, J.; Dietz, M.J.; Li, B. Antimicrobial peptide LL-37 is bactericidal against *Staphylococcus aureus* biofilms. *PLoS ONE* **2019**, *14*, e0216676. [CrossRef] [PubMed]
46. Segev-Zarko, L.-a.; Saar-Dover, R.; Brumfeld, V.; Mangoni, M.L.; Shai, Y. Mechanisms of biofilm inhibition and degradation by antimicrobial peptides. *Biochem. J.* **2015**, *468*, 259–270. [CrossRef] [PubMed]
47. Okuda, K.-I.; Zendo, T.; Sugimoto, S.; Iwase, T.; Tajima, A.; Yamada, S.; Sonomoto, K.; Mizunoe, Y. Effects of bacteriocins on methicillin-resistant *Staphylococcus aureus* biofilm. *Antimicrob. Agents Chemother.* **2013**, *57*, 5572–5579. [CrossRef] [PubMed]
48. Pulido, D.; Prats-Ejarque, G.; Villalba, C.; Albacar, M.; Gonzalez-Lopez, J.J.; Torrent, M.; Moussaoui, M.; Boix, E. A novel RNase 3/ECP peptide for *Pseudomonas aeruginosa* biofilm eradication that combines antimicrobial, lipopolysaccharide binding, and cell-agglutinating activities. *Antimicrob. Agents Chemother.* **2016**, *60*, 6313–6325. [CrossRef]
49. Zhang, Z.; Nadezhina, E.; Wilkinson, K.J. Quantifying diffusion in a biofilm of *Streptococcus mutans*. *Antimicrob. Agents Chemother.* **2011**, *55*, 1075–1081. [CrossRef]
50. Chen, C.Z.; Cooper, S.L. Interactions between dendrimer biocides and bacterial membranes. *Biomaterials* **2002**, *23*, 3359–3368. [CrossRef]
51. Huang, H.W. Action of antimicrobial peptides: Two-state model. *Biochemistry* **2000**, *39*, 8347–8352. [CrossRef]
52. Shai, Y. Mode of action of membrane active antimicrobial peptides. *Pept. Sci. Orig. Res. Biomol.* **2002**, *66*, 236–248. [CrossRef]
53. Mataraci, E.; Dosler, S. In vitro activities of antibiotics and antimicrobial cationic peptides alone and in combination against methicillin-resistant *Staphylococcus aureus* biofilms. *Antimicrob. Agents Chemother.* **2012**, *56*, 6366–6371. [CrossRef]
54. Cirioni, O.; Giacometti, A.; Ghiselli, R.; Kamysz, W.; Orlando, F.; Mocchegiani, F.; Silvestri, C.; Licci, A.; Chiodi, L.; Lukasiak, J.; et al. Citropin 1.1-treated central venous catheters improve the efficacy of hydrophobic antibiotics in the treatment of experimental staphylococcal catheter-related infection. *Peptides* **2006**, *27*, 1210–1216. [CrossRef]
55. Mohamed, M.F.; Abdelkhalek, A.; Seleem, M.N. Evaluation of short synthetic antimicrobial peptides for treatment of drug-resistant and intracellular *Staphylococcus aureus*. *Sci. Rep.* **2016**, *6*, 29707. [CrossRef]
56. Chung, P.Y.; Khanum, R. Antimicrobial peptides as potential anti-biofilm agents against multidrug-resistant bacteria. *J. Microbiol. Immunol. Infect.* **2017**, *50*, 405–410. [CrossRef] [PubMed]
57. Paduszynska, M.A.; Greber, K.E.; Paduszynski, W.; Sawicki, W.; Kamysz, W. Activity of temporin a and short lipopeptides combined with gentamicin against biofilm formed by *Staphylococcus aureus* and *Pseudomonas aeruginosa*. *Antibiotics* **2020**, *9*, 566. [CrossRef] [PubMed]
58. Behrendt, R.; White, P.; Offer, J. Advances in Fmoc solid-phase peptide synthesis. *J. Pep. Sci.* **2016**, *22*, 4–27. [CrossRef] [PubMed]
59. Gongora-Benítez, M.; Tulla-Puche, J.; Albericio, F. Handles for Fmoc Solid-Phase synthesis of protected peptides. *ACS Comb. Sci.* **2013**, *15*, 217–228. [CrossRef] [PubMed]
60. Yasir, M.; Dutta, D.; Kumar, N.; Willcox, M.D.P. Interaction of the surface bound antimicrobial peptides melimine and Mel4 with *Staphylococcus aureus*. *Biofouling* **2020**, *36*, 1019–1030. [CrossRef]
61. Wiegand, I.; Hilpert, K.; Hancock, R.E. Agar and broth dilution methods to determine the minimal inhibitory concentration (MIC) of antimicrobial substances. *Nat. Protoc.* **2008**, *3*, 163–175. [CrossRef]
62. Mishra, B.; Wang, G. Individual and combined effects of engineered peptides and antibiotics on *Pseudomonas aeruginosa* biofilms. *Pharmaceuticals* **2017**, *10*, 58. [CrossRef] [PubMed]

Article

Antimicrobial Efficacy of an Ultraviolet-C Device against Microorganisms Related to Contact Lens Adverse Events

Srikanth Dumpati, Shehzad A. Naroo, Sunil Shah and Debarun Dutta *

Optometry School, College of Health and Life Sciences, Aston University, Birmingham B4 7ET, UK; 200285764@aston.ac.uk (S.D.); s.a.naroo@aston.ac.uk (S.A.N.); s.shah26@aston.ac.uk (S.S.)
* Correspondence: d.dutta@aston.ac.uk; Tel.: +44-121-204-4418

Abstract: The purpose of the study was to assess the antimicrobial activity of an ultraviolet-C (UVC) device against microorganisms implicated in contact lens related adverse events. An UVC device with an emitting 4.5 mm diameter Light Emitting Diode (LED; 265 nm; 1.93 mJ/cm^2) was used. *Pseudomonas aeruginosa*, *Staphylococcus aureus*, *Fusarium solani*, and *Candida albicans* agar plate lawns were exposed to the device beams for 15 and 30 s at 8 mm distance. Following the exposure, the diameter of the growth inhibition zone was recorded. Contact lenses made of Delfilcon-A, Senofilcon-A, Comfilcon-A, Balafilcon-A, Samfilcon-A and Omafilcon-A and a commercially available contact storage case was used. They were exposed to bacterial and fungal strains for 18 h at 37 °C and 25 °C respectively. After this, the samples were exposed to UVC for 30 s at 8 mm distance to determine the antimicrobial efficacy. Samples were then gently washed and plated on appropriate agar for enumeration of colonies. The UVC exposure reduced microbial growth by 100% in agar lawns, and significantly ($p < 0.05$) reduced microbial contamination to contact lenses and cases, ranging between 0.90 to 4.6 log. Very short UVC exposure has high antimicrobial efficacy against most of the predominant causative microorganisms implicated in contact lens related keratitis. UVC could be readily used as a broad-spectrum antimicrobial treatment for lens disinfection.

Keywords: contact lenses; ultraviolet C; keratitis; Pseudomonas; Staphylococcus; Fusarium; Candida; antibiotic resistance

1. Introduction

Contact lenses are an increasingly popular option for refractive correction with current estimates of more than 140 million wearers worldwide [1]. In addition, contact lenses are indispensable for patients with high astigmatism, high refractive error, irregular astigmatism, myopia control and are regularly used for post-surgical therapeutic use. However, contact lenses can be associated with various microbial adverse events such as microbial keratitis (MK), contact lens acute red eye (CLARE), contact lens peripheral ulcer (CLPU) and infiltrative keratitis (IK) [2].

MK is a worldwide medical concern often noted as the most serious form of contact lens infection, in the UK, 65% of all new cases of MK are contact lens-related [3]. The incidence of contact lens related-MK is around 4 per 10,000 a year for daily wear and 20 per 10,000 a year for extended wear [4]. Other less severe conditions have an even higher incidence whereby CLARE has been found to occur in up to 34% of those who regularly wear hydrogel contact lenses [5]. Sixty-six percent of complications observed in contact lens wearers are attributed to poor handling of lenses and lens cases [6]. Despite the introduction of silicone hydrogel materials, advancement in care products and cleaning regimens, the incidence of contact lens-related microbial adverse events remained unchanged [7]. The emergence of antibiotic and preservative resistant opportunistic microorganisms has further complicated the treatment options. It is well known that MK caused by antibiotic-resistant microorganisms are associated with longer hospitalization and poorer visual outcome [8].

There is a great need for an alternative antimicrobial strategy for millions of lens wearers worldwide that may provide broad-spectrum antimicrobial activity bypassing our reliance on preservatives and antibiotic use.

Ultraviolet light (UV) is part of the electromagnetic spectrum and can be divided into four distinct spectral areas: UVA (wavelengths 315–400 nm); UVB (wavelengths 280–315 nm); UVC (wavelengths 200–280 nm); and vacuum UV (wavelengths 100–200 nm) [9]. Amongst these wavelength ranges, UVC has the highest capacity to inactivate microorganisms because the peak germicidal wavelength is in the range of 250–270 nm and is known as the germicidal spectrum [10]. UVC cause cellular damage by inducing changes in the chemical structure of DNA chains [11]. The consequence is the production of cyclobutene pyrimidine dimers (CPDs) causing distortion of the DNA molecule, which may cause malfunctions in cell replication and lead to cell death [9]. Effect of UVC treatment on sterilization of contact lenses and cases have been reported before [12,13]. UVC has been shown to have high efficacy in killing acanthamoeba cysts following exposure up to 24 minutes [14]. Attempts have been made to incorporate UVC within contact lens disinfection systems which showed statistically significant reduction in microbial load [15,16].

UVC irradiation is well known for its germicidal action, however, the use of UVC irradiation for prevention and treatment of localized infections is still in the early stages of development. Previous studies confirmed that UVC inactivation is equally effective to antibiotic-resistant bacteria compared to their native counterparts [17].

This study aimed to investigate the antimicrobial activity of UVC against major microorganisms related to contact lens-related keratitis. A further aim was to determine the potential application of UVC in reducing the microbial contamination of contact lenses and lens cases.

2. Results

UVC device showed very high antimicrobial activity against all the microorganisms tested. When tested with contact lenses and lens cases, the UVC device showed a significant reduction in contamination most of the time.

Figure 1 shows inhibition zones of microbial agar lawns following exposure to UVC for 15 and 30 s. The areas exposed to UVC showed inhibition zones, rest of the control areas showed confluent bacterial growth. Both 15 and 30 s exposure were able to fully inhibit microbial growth as identified by the arrow in Figure 1. The diameter of the inhibition zone was slightly increased for 30 s compared to 15 s (Table 1).

Table 1. Inhibition zone diameter (mm) following UVC exposure.

Microorganisms	15 s Exposure	30 s Exposure
P. aeruginosa 6294	7.2 ± 0.3	7.5 ± 0.4
S. aureus 38	6.9 ± 0.3	7.1 ± 0.4
C. albicans ATCC 76615	5.5 ± 0.4	5.9 ± 0.2
F. solani ATCC 10696	5.0 ± 0.4	5.8 ± 0.5

The following Table 1 demonstrates that all the microorganisms showed complete inhibition zones, and the area of inhibition varied between the microorganisms tested. *P. aeruginosa* lawns had the largest inhibition zones compared to other microorganisms, whereas *F. solani* had the smallest. Exposure of 30 s had a slightly larger inhibition zone compared to 15 s exposure. This difference was highest with *F. solani* (0.8 ± 0.2 mm), and lowest for *S. aureus* (0.2 ± 0.1 mm).

The antimicrobial efficacy of UVC treatment on various contact lens materials and lens cases contaminated with *P. aeruginosa* is detailed in Figure 2. Significant ($p < 0.001$) reduction of *P. aeruginosa* contamination was noted following exposure to UVC for all contact lens materials. Reduction of contamination for lens case was $36.2 \pm 13.3\%$ (0.28 ± 0.09 log; $p = 0.194$).

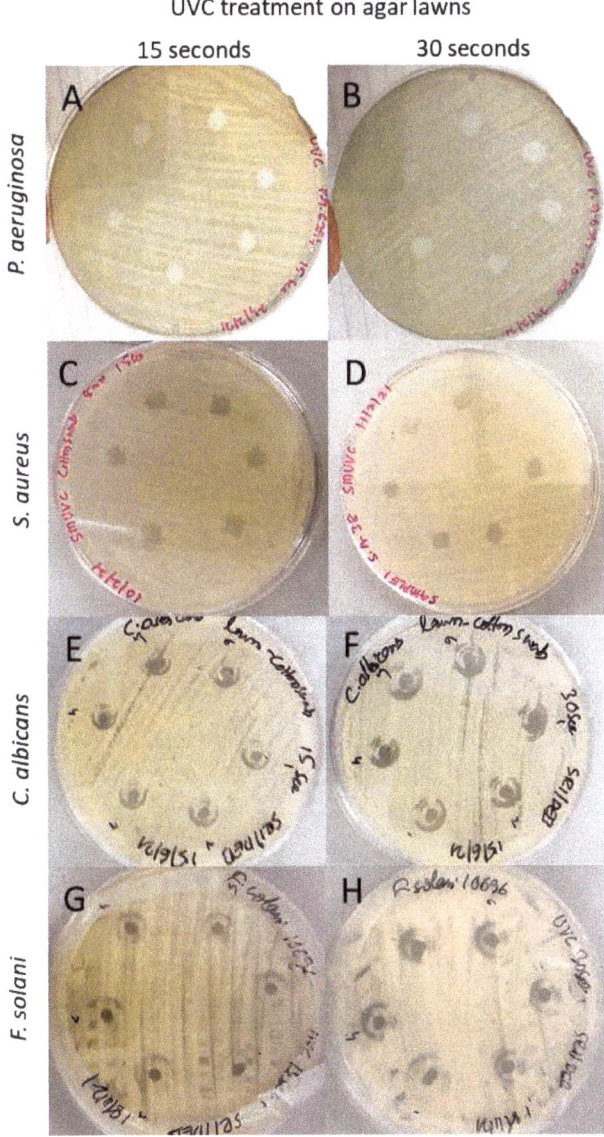

Figure 1. Representative photographs of the agar plates following 15 and 30 s of exposure to UVC device. The photographs demonstrate that 15 and 30 s exposure areas of complete growth inhibition of (**A,B**) *P. aeruginosa*, (**C,D**) *S. aureus*, (**E,F**) *C. albicans*, (**G,H**) *F. solani*.

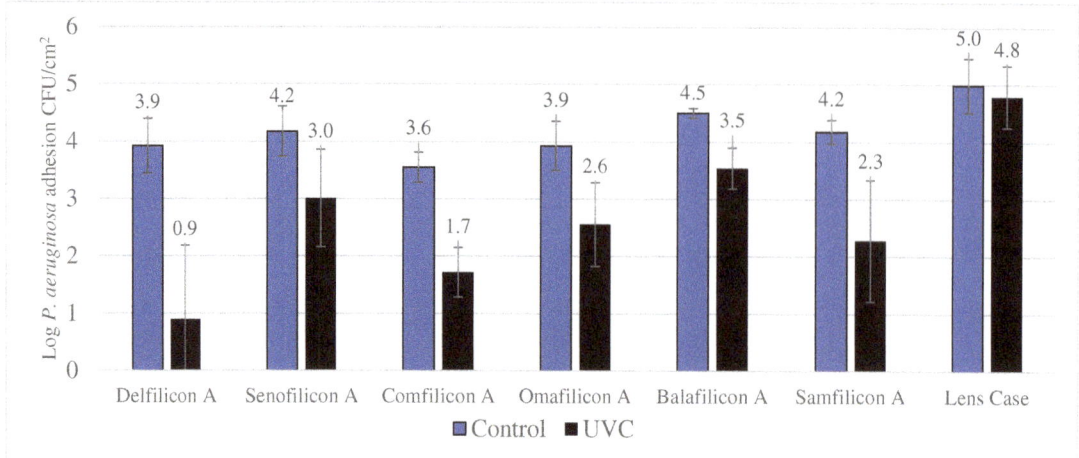

Figure 2. Reduction of *P. aeruginosa* contamination after UVC treatment. Exposure to UVC statistically significant ($p < 0.001$) reduced *P. aeruginosa* contamination of contact lenses. The reduction in contamination observed with lens case showed no significant difference ($p = 0.194$).

The efficacy of UVC treatment on various contact lens materials and lens cases contaminated with *S. aureus* is detailed in Figure 3. Significant ($p < 0.001$) reduction in contamination was observed against all the tested contact lens materials and the lens case.

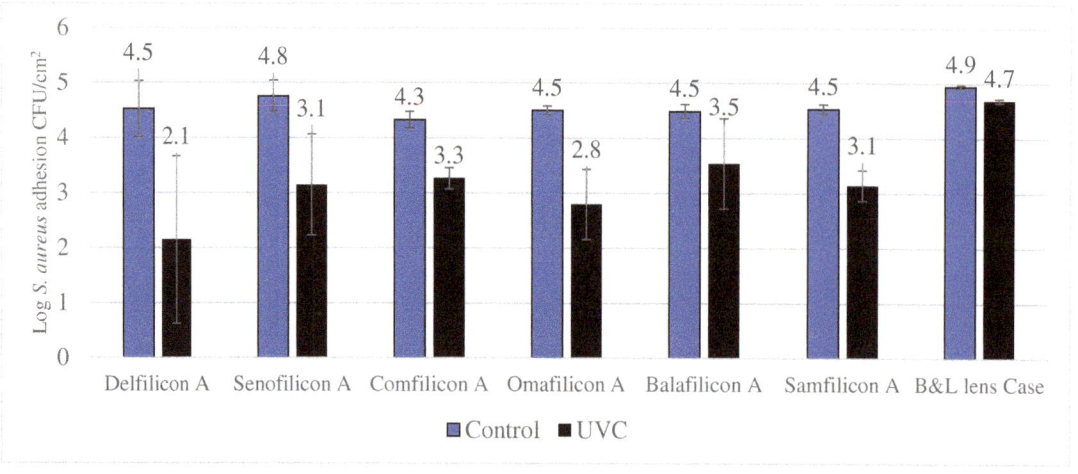

Figure 3. Reduction of *S. aureus* contamination following UVC treatment. Exposure to UVC statistically significant ($p < 0.001$) reduced *S. aureus* contamination of contact lenses and lens cases.

Reduction of *C. albicans* contamination in contact lenses and lens case following exposure to UVC is detailed in Figure 4. Significant ($p < 0.001$) reduction in contamination was observed against all the tested contact lens materials and the lens case.

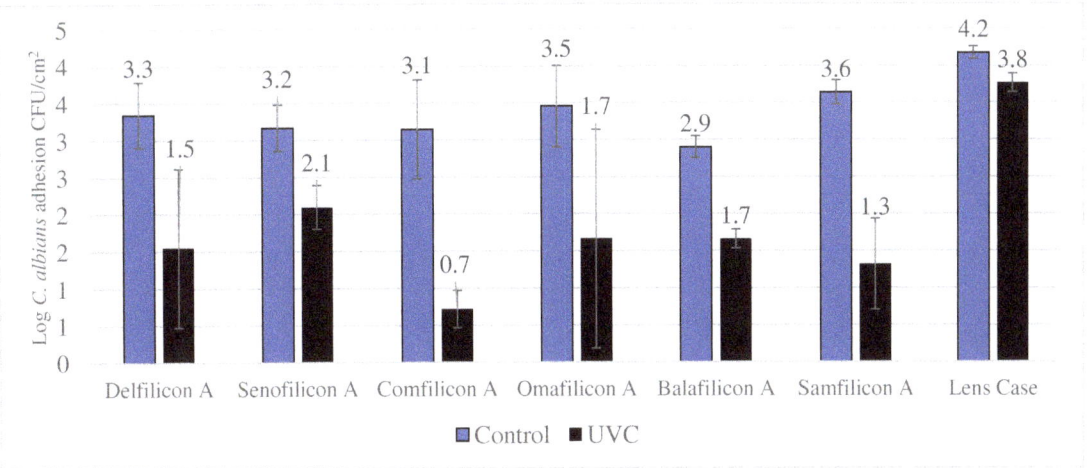

Figure 4. Reduction of *C. albicans* contamination after UVC treatment. Exposure to UVC statistically significant ($p < 0.001$) reduced *C. albicans* contamination of all contact lenses and lens cases.

The efficacy of UVC exposure to *F. solani* contaminated lenses and lens cases are demonstrated in Figure 5. Overall high efficacy in reduction of contamination was observed for all contact lens materials except for Balafilcon-A (0.55 ± 0.13 log; $p = 0.189$) and Samfilcon-A (0.70 ± 0.26 log; $p = 0.110$). The antimicrobial efficacy with lens case was $90.4 \pm 3.3\%$ (1.02 ± 0.39 log) which was statistically significant ($p = 0.001$).

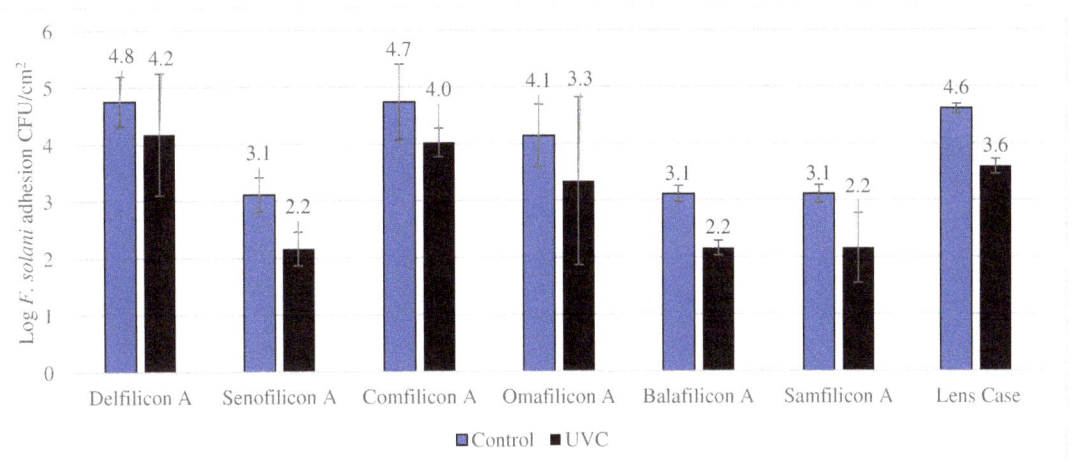

Figure 5. Reduction of *F. solani* contamination after UVC treatment. Exposure to UVC statistically significant ($p < 0.05$) reduced *F. solani* contamination of Delefilcon-A, Senofilcon-A, Comfilcon-A, and Samfilcon-A contact lens materials and lens cases.

The following Table 2. summarizes the reduction of the percentage of contact lens contamination implicated by UVC treatment.

Table 2. Percent of reduction in contact lens microbial contamination (mean ± SD) following UVC treatment. Asterix (*) indicates statistically significant difference.

Contact Lenses	P. aeruginosa	S. aureus	C. albicans	F. solani
Delefilcon A	99.9 ± 5.2 *	99.6 ± 10.3 *	98.4 ± 26.7 *	76.0 ± 5.3 *
Senofilcon A	93.2 ± 4.3 *	97.5 ± 8.9 *	91.5 ± 13.3 *	68.3 ± 7.3 *
Comfilcon A	98.5 ± 14.3 *	91.2 ± 5.8 *	99.6 ± 4.2 *	80.5 ± 13.1 *
Omafilcon A	97.6 ± 15.5 *	98.0 ± 8.5 *	98.4 ± 17.8 *	73.7 ± 8.6 *
Balafilcon A	89.0 ± 7.8 *	88.8 ± 13.3 *	94.3 ± 3.4 *	71.5 ± 10.3
Samfilcon A	98.7 ± 7.1 *	95.9 ± 7.8 *	99.5 ± 10.3 *	79.9 ± 26.3
Lens Case	36.7 ± 13.3	44.7 ± 12.6 *	61.2 ± 3.4 *	90.4 ± 3.3 *

3. Discussion

The current study found that a very short 15–30 s exposure of UVC can provide high antimicrobial action against most of the predominant microorganisms responsible for contact lens keratitis. In addition, this treatment can substantially reduce contact lens and lens case contamination, with a real potential to reduce these types of keratitis in a clinical setting.

The UVC device showed total efficacy against P. aeruginosa, S. aureus, F. solani, and C. albicans when exposed directly to agar lawns. The 4.5 mm UVC exposure to the microbial lawns showed 5.0 to 5.9 mm inhibition against the fungal strains and 6.9 mm to 7.5 mm inhibition zone against the bacterial strains. Inhibition zones were bigger with bacterial strains compared to fungal strains, which may be because the bacteria at the edges of the inhibition zones were more sensitive to UVC compared to fungal strains, which is supported by the contact lens contamination study where inhibition on bacterial strains was higher compared to fungal strains. The results reported in the current study are slightly higher than previously reported by Dean et al. [18], which showed 3.50 mm to 5.50 mm inhibition zone against bacterial strains, however they did not check against fungal strains. Thai et al. used 254 nm UVC and showed that 180 s of exposure can significantly reduce bacterial load on chronic wounds [19]. Guridi et al. used varying doses (840–3360 mJ/cm^2) of UVC (253.7 nm) against P. aeruginosa, S. aureus, and C. albicans and found >99.99% efficacy when exposed directly on different biomaterial surfaces [20]. This is in agreement with our results on direct exposure, including on bacteria on contact lens surfaces which often showed >99% reduction in bacterial viability. Umezawa et al. investigated the efficacy of pulsed UVC light (photon peaks spread across 240–400 nm), which showed more than 2 log growth inhibition against similar microorganisms such as P. aeruginosa and S. aureus [21]. The efficacy of UVC (254 nm) against similar microorganisms on textile surfaces are reported to be more than 90% [22], which is also in line with our reports with high antimicrobial efficacy.

Contamination of contact lenses and lens cases have been directly implicated in the development of corneal infiltrative events, particularly in various types of keratitis [2,23]. Several antimicrobial strategies have been adopted to reduce contamination of causative microorganisms such as Gram-negative and Gram-positive bacteria and fungal strains [2,24]. Preservatives and disinfectants are the first-line antimicrobial agents used in contact lens care solutions. However more than 50% of lens cases from asymptomatic lens wearers were found to be contaminated, and more than 10% were with opportunistic Gram-negative bacteria [25]. Silver, selenium, Salicylic acid, Fimbrolides and antimicrobial peptides are some of the common strategies that were investigated as additional antimicrobials in the past [2,24,26,27]. The current study indicated that combining UVC treatment with existing care regimes is likely to significantly reduce contamination levels.

Previous studies have shown that the rate of microbial contamination can significantly vary based on the type of contact lens material used, while 2nd generation silicone hydrogel lenses may attract more microorganisms compared to hydrogel lenses [28]. However, the current study found bacterial contamination to the variety of control lenses are comparable. UVC exposure was able to significantly reduce contamination of both *P. aeruginosa* and *S. aureus* for silicone hydrogel and hydrogel lenses. The inhibition against *P. aeruginosa* ranged between 89% to more than 99% whereas against *S. aureus* inhibition ranged between 88% to 99%. There was no particular pattern found between different types of silicone hydrogel lenses, whereas Delefilcon A lenses were associated with the highest antimicrobial efficacy; more than 2.5 log inhibition with UVC against *P. aeruginosa*. Similar results were observed with *S. aureus*.

Depending on the geographical location, contact lens wear is often the most common risk factor for the development of fungal keratitis. This can often exceed 50% of the cases. Fusarium and Candida are the most common types of fungal strains implicated in contact lens-related keratitis, isolated from 41% and 14% of culture-positive tests [29]. The current study showed that UVC irradiation can significantly reduce *C. albicans* contamination ranging between 1.07 to 2.43 log inhibition based on the type of contact lens material used. A similar trend was observed against *F. solani*, where UVC showed inhibition ranging between 0.90 to 0.71 log. Although UVC showed 71% and 79% inhibition against *F. solani* in Balafilcon-A and Samfilcon-A lens materials, the differences were not statistically significant. The current study is one of the few studies that has used Balafilcon-A and it is the first study to have used Samfilcon-A for investigation with fungal strains, thus requiring further investigation with these lens materials. The overall, high antifungal efficacy of UVC irradiation coupled with the existing contact lens care regimen would certainly provide high and comprehensive fungicidal activity, protective towards contact lens-related fungal keratitis.

This study found that microbial contamination of contact lens cases was higher compared to lenses, which is likely due to the formation of biofilms of lens cases. Various enzymes, antibiofilm peptides, and other dispersion molecules have been investigated for medical-biofilm dispersion [30]. A limited number of agents including antimicrobial peptides, furanones, silver and passive dispersion agents have been tested on contact lens cases [2]. Contact lens cases are known to harbour microbial biofilms and have been directly associated with keratitis events [31]. The current study showed that UVC irradiation can significantly reduce *S.aureus* (44%), *C. albicans* (61%) and *F. solani* (90%), however, only 36% inhibition against *P. aeruginosa* was achieved.

This study did not investigate the safety of the UVC which was reported earlier [18]. Dean et al. reported that up to 30 s of exposure to UVC did not stimulate the death of human corneal epithelium [18]. Although the current study did not expose human cells to UVC, it is important to note that UVC has very little penetration on the cornea and is unlikely to impact the corneal endothelium. UV rays are known to cause photokeratitis which is also called ultraviolet photokeratitis. However, the ocular tissue damage threshold for UV rays is 5 mJ/cm^2, and the LED used for irradiation in this study emits less than 2 mJ/cm^2. Given that the UVC exposure is aimed to decontaminate contact lenses and lens cases, accidental exposure to the eye is unlikely to cause any major harm. This study did not examine any detrimental effect of direct UVC exposure to contact lenses. Polymerization of contact lens monomer include exposure to UVC, hence we assumed that the short exposure of UVC to contact lens materials unlikely to have any significant change in the key parameters such as base curve, diameter, refractive index and oxygen transmissibility.

4. Materials and Methods

4.1. Ultraviolet C device:

The prototype device comprises a 265-nm (Figure 6) (Photon Therapeutics; Oldsmar, UK) detailed earlier [18]. Briefly, it contains a hemispheric ball lens, which is protected by a rubber sheath 8 mm length, projecting a spot size of 4.5 mm, resulting in an intensity of 1.93 mJ/cm^2 at the target distance, as confirmed with a calibrated UVC light meter (Solar meter Model 8.0 UVC, Solartech Inc, Harrison Twp, MI, USA). Power was supplied by a 9 V DC regulated adapter with an additional current limiting circuit [18].

Figure 6. UVC device in this study.

Bacterial lawns were freshly prepared on Nutrient Agar (NA; Sigma Aldrich, St. louis, MO, USA) and fungal lawns were made on Potato Dextrose Agar (PDA: Merck Ga A, Damstadt, Germany) plates from the previously prepared suspensions. The plates were exposed to 4.5 mm diameter UVC beam for 15 and 30 s at an 8 mm distance. After 24 h incubation at 37 °C for bacteria or 2 days incubation at 37 °C for *C. albicans* and 4 days incubation at 25 °C for *F. solani*, the efficacy of the UVC beam was examined by investigating the diameter of the treatment zone, using a digital colony counter (Stuart Company, London, UK). A total of three horizontal and three vertical measurement of the inhibition zone were made and the average and standard deviation was reported.

4.2. Contact lenses and Lens cases

Widely used and most popular contact lenses were used in this study, their parameters, materials and other properties are described in Table 3. Bausch and Lomb contact lens cases (Bausch and Lomb UK Ltd., Kingston, UK) were used in this study.

Table 3. Properties of contact lens materials used in the study.

Proprietary Name	Total Dailes1	Acuvue Oasys	Biofinity	Proclear	Purevision2	Ultra
United States Adopted Name (USAN)	Delfilcon A	Senofilcon A	Comfilcon A	Omafilcon A	Balafilcon A	Samfilcon A
Lens material	Silicone hydrogel	Silicone hydrogel	Silicone hydrogel	Hydrogel	Silicone Hydrogel	Silicone Hydrogel
Manufacturer	Alcon	Johnson & Johnson	Cooper vision	Cooper vision	Bausch & Lomb	Bausch & Lomb
Water content (%)	Gradient	38	48	62	36	46
Oxygen transmissibility (DK/t)	156	147	160	37	130	163
Centre thickness (mm) -3.00DS	0.09 mm	0.07 mm	0.08 mm	0.09 mm	0.07 mm	0.07 mm

4.3. Strains and microbial conditions

Pseudomonas aeruginosa strain 6294 and *Staphylococcus aureus* strain 38 isolated from MK cases were used in this study. *Fusarium solani* ATCC 10696 isolated from soil and *Candida albicans* ATCC 76615, a clinical isolate were used in this study. Bacteria were grown overnight in TSB (Melfold, UK) at 37 °C with aeration. The harvested bacterial cells were centrifuged for 10 minutes at 3000 rpm and the cells were washed three times with phosphate-buffered saline (PBS; pH 7.4; NaCl 8 g L^{-1}, KCl 0.2 g L^{-1}, Na_2HPO_4 1.15 g L^{-1}, KH_2PO_4 0.2 g L^{-1}). *P. aeruginosa* were then resuspended in PBS and *S. aureus* were resuspended in 10% TSB to an OD$_{660nm}$ of 0.1 (1×10^8 CFU mL^{-1}). The bacterial cell suspensions were then diluted to 1×10^6 CFU mL^{-1}. *C. albicans* strains were grown on PDA plates by incubating for 24 h at 37 °C, then suspended in sterile PBS to an OD$_{660nm}$ of 1.5 (1×10^8 CFU mL^{-1}) and the suspensions were serially diluted to 1.0×10^6 CFU mL^{-1} and used for adhesion assays. *F. solani* were grown on PDA plates by incubating for 7 to 10 days at 25 °C followed by filtering through sterile 70 μm filters to remove hyphal fragments and finally resuspended to an OD$_{660nm}$ of 2.6 (1×10^8 CFU mL^{-1}).

Microbial assays with contact lenses and lens cases have been detailed earlier [32]. Briefly, contact lenses were washed two times in PBS and transferred to 1ml of bacterial or fungal suspensions in wells of 24-well tissue culture plates (CELESTAR®, Greiner bio-one, Frickenhausen, Germany), keeping concave side up. To allow contamination, lenses were incubated with 1mL bacterial suspension for 18 h at 37 °C and for fungal strains 18 h at 25 °C with shaking (120 rpm). Lens cases were incubated similarly but with 2 mL microbial suspensions in the lens case cup. After this, lenses were aseptically removed from the microbial suspensions and washed twice with 1ml PBS in a 24 well plate by shaking at 120 rpm for 30 s to remove non-adherent cells. Lens cases were washed with 1 mL PBS twice by shaking 120 rpm for 30 s.

Following exposure to microorganisms, each contact lens was cut into equal 4 samples with a sterile scalpel, one piece used as control and the rest three pieces were placed 8 mm beneath a 265 nm UVC lamp for 30 s. Four 4 mm non-overlapping UVC beams were exposed to both sides of the lens. Similarly, each lens case was exposed to 9 non-overlapping 30 s spots.

After this, all lens samples were placed in a 2 mL sterile plastic vial containing 2 mL PBS with a sterile magnetic bar and vortexed for at least one minute. Control and UVC-exposed lens cases were filled with 2 mL PBS and a sterile magnetic bar and vortexed for at least one minute. For bacterial and *C. albicans* strains, following log serial dilutions in PBS, three 50 micro-litre droplets of each dilution were plated on NA and PDA plates for recovery of cells respectively. For *F. solani*, following log serial dilutions in PBS, 100 micro-litre were plated onto PDA for recovery of viable cells. After 24 h incubation at 37 °C for bacteria or 2 days incubation at 37 °C for *C. albicans* and 4 days incubation at 25 °C for *F. solani*, the viable micro-organisms were enumerated as colony-forming units (CFU). Results are expressed as the reduction in viable bacteria or fungi (compared with the untreated control samples). Three samples were used for each experiment and were repeated for at least three separate occasions.

The adhesion data were $\log_{10}(x+1)$ transformed prior to data analysis where x is the adherent bacteria or fungi in colony-forming units. The reuction of adhesion data was presented as mean ± standard deviation. Differences in the microbial load were analyzed using the Wilcoxon-Signed ranked test. Differences between the groups were analyzed using linear mixed model ANOVA, which adjusts the correlation due to repeated observations. Post hoc multiple comparisons were done using Bonferroni correction. Statistical significance was set at 5%.

5. Conclusions

In conclusion, this study showed that the ophthalmic device with a very short UVC exposure has potent antimicrobial activity against a majority of the causative microorganisms for contact lens-related keratitis. The device is particularly effective in reducing contamination on contact lenses and lens cases.. This study further demonstrate that UVC could be readily used as a preventative measure and inhibition of broad-spectrum antimicrobial contamination.

Author Contributions: Conceptualization, D.D., S.A.N. and S.S.; methodology, D.D., S.A.N. and S.S.; investigation, S.D.; resources, D.D. and S.S.; data curation, S.D.; writing—original draft preparation, S.D.; writing—review and editing, D.D., S.A.N. and S.S.; supervision, D.D., S.A.N. and S.S.; project administration, D.D.; funding acquisition, D.D. and S.S. All authors have read and agreed to the published version of the manuscript.

Funding: This research was funded by Marie Skłodowska-Curie COFUND Programme of European Union, grant number 713694 and by Photon Therapeutics Ltd.

Institutional Review Board Statement: Not Applicable.

Informed Consent Statement: Not applicable.

Data Availability Statement: This study did not report any data.

Acknowledgments: The authors thank Mark Willcox of the University of New South Wales, and Rachel Williams of Liverpool University for helping with the microbial strains used in this study.

Conflicts of Interest: S.S. is co-founder of Photon Therapeutics Ltd., a company that holds a patent in the use of thera-peutic UVC.

References

1. Stellwagen, A.; MacGregor, C.; Kung, R.; Konstantopoulos, A.; Hossain, P. Personal hygiene risk factors for contact lens-related microbial keratitis. *BMJ Open Ophthalmol.* **2020**, *5*, e000476. [CrossRef] [PubMed]
2. Dutta, D.; Willcox, M.D.P. Antimicrobial contact lenses and lens cases: A review. *Eye Contact Lens* **2014**, *40*, 312–324. [CrossRef] [PubMed]
3. Hu, X.; Shi, G.; Liu, H.; Jiang, X.; Deng, J.; Zhu, C.; Yuan, Y.; Ke, B. Microbial Contamination of Rigid Gas Permeable (RGP) Trial Lenses and Lens Cases in China. *Curr. Eye Res.* **2020**, *45*, 550–555. [CrossRef]
4. Stapleton, F.; Keay, L.; Edwards, K.; Naduvilath, T.; Dart, J.K.; Brian, G.; Holden, B.A. The Incidence of Contact Lens–Related Microbial Keratitis in Australia. *Ophthalmology* **2008**, *115*, 1655–1662. [CrossRef] [PubMed]
5. Sankaridurg, P.R.; Sharma, S.; Willcox, M.; Naduvilath, T.J.; Sweeney, D.F.; Holden, B.A.; Rao, G.N. Bacterial colonization of disposable soft contact lenses is greater during corneal infiltrative events than during asymptomatic extended lens wear. *J. Clin. Microbiol.* **2000**, *38*, 4420–4424. [CrossRef]
6. Brewitt, H. Contact lenses. Infections and hygiene. *Ophthalmologe* **1997**, *94*, 311–316. [CrossRef]
7. Chalmers, R.L.; Hickson-Curran, S.B.; Keay, L.; Gleason, W.J.; Albright, R. Rates of Adverse Events With Hydrogel and Silicone Hydrogel Daily Disposable Lenses in a Large Postmarket Surveillance Registry: The TEMPO Registry. *Investig. Opthalmol. Vis. Sci.* **2015**, *56*, 654–663. [CrossRef]
8. Ting, D.S.J.; Ho, C.S.; Deshmukh, R.; Said, D.G.; Dua, H.S. Infectious keratitis: An update on epidemiology, causative microorganisms, risk factors, and antimicrobial resistance. *Eye* **2021**, *35*, 1084–1101. [CrossRef]
9. Dai, T.; Vrahas, M.S.; Murray, C.K.; Hamblin, M.R. Ultraviolet C irradiation: An alternative antimicrobial approach to localized infections? *Expert Rev. Anti-Infect. Ther.* **2012**, *10*, 185–195. [CrossRef]
10. Gurzadyan, G.G.; Görner, H.; Schulte-Frohlinde, D. Ultraviolet (193, 216 and 254 nm) Photoinactivation of *Escherichia coli* Strains with Different Repair Deficiencies. *Radiat. Res.* **1995**, *141*, 244. [CrossRef]
11. Chang, J.C.; Ossoff, S.F.; Lobe, D.C.; Dorfman, M.H.; Dumais, C.M.; Qualls, R.G.; Johnson, J.D. UV inactivation of pathogenic and indicator microorganisms. *Appl. Environ. Microbiol.* **1985**, *49*, 1361–1365. [CrossRef] [PubMed]
12. Gritz, D.C.; Lee, T.Y.; McDonnell, P.J.; Shih, K.; Baron, N. Ultraviolet radiation for the sterilization of contact lenses. *CLAO J.* **1990**, *16*, 294–298. [PubMed]
13. Harris, M.G.; Fluss, L.; Lem, A.; Leong, H. Ultraviolet Disinfection of Contact Lenses. *Optom. Vis. Sci.* **1993**, *70*, 839–842. [CrossRef] [PubMed]
14. Lonnen, J.; Putt, K.S.; Kernick, E.R.; Lakkis, C.; May, L.; Pugh, R.B. The Efficacy of Acanthamoeba Cyst Kill and Effects Upon Contact Lenses of a Novel Ultraviolet Lens Disinfection System. *Am. J. Ophthalmol.* **2014**, *158*, 460–468.e2. [CrossRef]
15. Admoni, M.M.; Bartolomei, A.; Qureshi, M.N.; Bottone, E.J.; Asbell, P.A. Disinfection Efficacy in an Integrated Ultraviolet Light Contact Lens Care System. *Eye Contact Lens Sci. Clin. Pract.* **1994**, *20*, 246–248. [CrossRef]
16. Choate, W.; Fontana, F.; Potter, J.; Schachet, J.; Shaw, R.; Soulsby, M.; White, E. Evaluation of the PuriLens contact lens care system: An automatic care system incorporating UV disinfection and hydrodynamic shear cleaning. *CLAO J.* **2000**, *26*, 134–140.
17. Conner-Kerr, T.; Sullivan, P.K.; Gaillard, J.; Franklin, M.; Jones, R.M. The effects of ultraviolet radiation on antibiotic-resistant bacteria in vitro. *Ostomy Wound Manag.* **1998**, *44*, 50–56.
18. Dean, S.J.; Petty, A.; Swift, S.; McGhee, J.J.; Sharma, A.; Shah, S.; Craig, J.P. Efficacy and safety assessment of a novel ultraviolet C device for treating corneal bacterial infections. *Clin. Exp. Ophthalmol.* **2011**, *39*, 156–163. [CrossRef]
19. Thai, T.P.; Keast, D.; Campbell, K.; Woodbury, M.G.; Houghton, P. Effect of ultraviolet light C on bacterial colonization in chronic wounds. *Ostomy Wound Manag.* **2005**, *51*, 32–45.
20. Guridi, A.; Sevillano, E.; de la Fuente, I.; Mateo, E.; Eraso, E.; Quindós, G. Disinfectant Activity of A Portable Ultraviolet C Equipment. *Int. J. Environ. Res. Public Health* **2019**, *16*, 4747. [CrossRef]
21. Umezawa, K.; Asai, S.; Inokuchi, S.; Miyachi, H. A Comparative Study of the Bactericidal Activity and Daily Disinfection Housekeeping Surfaces by a New Portable Pulsed UV Radiation Device. *Curr. Microbiol.* **2012**, *64*, 581–587. [CrossRef]
22. Bentley, J.J.; Santoro, D.; Gram, D.W.; Dujowich, M.; Marsella, R. Can ultraviolet light C decrease the environmental burden of antimicrobial-resistant and -sensitive bacteria on textiles? *Vet. Dermatol.* **2016**, *27*, 457-e121. [CrossRef] [PubMed]

23. Dutta, D.; Cole, N.; Willcox, M. Factors influencing bacterial adhesion to contact lenses. *Mol. Vis.* **2012**, *18*, 14–21.
24. Willcox, M.D.P.; Chen, R.; Kalaiselvan, P.; Yasir, M.; Rasul, R.; Kumar, N.; Dutta, D. The Development of an Antimicrobial Contact Lens—From the Laboratory to the Clinic. *Curr. Protein Pept. Sci.* **2020**, *21*, 357–368. [CrossRef] [PubMed]
25. Szczotka-Flynn, L.B.; Pearlman, E.; Ghannoum, M. Microbial Contamination of Contact Lenses, Lens Care Solutions, and Their Accessories: A Literature Review. *Eye Contact Lens Sci. Clin. Pract.* **2010**, *36*, 116–129. [CrossRef] [PubMed]
26. Yasir, M.; Dutta, D.; Willcox, M.D. Activity of Antimicrobial Peptides and Ciprofloxacin against *Pseudomonas aeruginosa* Biofilms. *Molecules* **2020**, *25*, 3843. [CrossRef] [PubMed]
27. Casciaro, B.; Dutta, D.; Loffredo, M.R.; Marcheggiani, S.; McDermott, A.M.; Willcox, M.D.; Mangoni, M.L. Esculentin-1a derived peptides kill Pseudomonas aeruginosa biofilm on soft contact lenses and retain antibacterial activity upon immobilization to the lens surface. *Biopolymers* **2017**. [CrossRef]
28. Dutta, D.; Willcox, M.D. A Laboratory Assessment of Factors That Affect Bacterial Adhesion to Contact Lenses. *Biology* **2013**, *2*, 1268–1281. [CrossRef]
29. Iyer, S.A.; Tuli, S.S.; Wagoner, R.C. Fungal Keratitis: Emerging Trends and Treatment Outcomes. *Eye Contact Lens Sci. Clin. Pract.* **2006**, *32*, 267–271. [CrossRef]
30. Fleming, D.; Rumbaugh, K.P. Approaches to Dispersing Medical Biofilms. *Microorganisms* **2017**, *5*, 15. [CrossRef]
31. Wu, Y.T.-Y.; Willcox, M.; Zhu, H.; Stapleton, F. Contact lens hygiene compliance and lens case contamination: A review. *Contact Lens Anterior Eye* **2015**, *38*, 307–316. [CrossRef] [PubMed]
32. Dutta, D.; Cole, N.; Kumar, N.; Willcox, M. Broad Spectrum Antimicrobial Activity of Melimine Covalently Bound to Contact Lenses. *Investig. Opthalmol. Vis. Sci.* **2013**, *54*, 175–182. [CrossRef] [PubMed]

Article

Biocompatibility and Comfort during Extended Wear of Mel4 Peptide-Coated Antimicrobial Contact Lenses

Parthasarathi Kalaiselvan [1,*], Debarun Dutta [1,2], Nagaraju Konda [3], Pravin Krishna Vaddavalli [4,5], Savitri Sharma [6], Fiona Stapleton [1] and Mark D. P. Willcox [1]

[1] School of Optometry and Vision Science, UNSW Sydney, Sydney, NSW 2041, Australia; d.dutta@aston.ac.uk (D.D.); f.stapleton@unsw.edu.au (F.S.); m.willcox@unsw.edu.au (M.D.P.W.)
[2] School of Optometry, Aston University, Birmingham B4 7ET, UK
[3] School of Medical Sciences, University of Hyderabad, Hyderabad 500 046, India; knr@uohyd.ac.in
[4] Bausch & Lomb Contact Lens Centre, L V Prasad Eye Institute, Hyderabad 500 034, India; pravin@lvpei.org
[5] The Cornea Institute, L V Prasad Eye Institute, Hyderabad 500 034, India
[6] Jhaveri Microbiology Centre, L V Prasad Eye Institute, Hyderabad 500 034, India; savitri@lvpei.org
* Correspondence: p.kalaiselvan@unsw.edu.au

Abstract: (1) Purpose: This study aimed to investigate the effects of Mel4 antimicrobial contact lenses (MACL) on the ocular surface and comfort during extended wear. (2) Methods: A prospective, randomised, double-masked, contralateral clinical trial was conducted with 176 subjects to evaluate the biocompatibility of contralateral wear of MACL. The wearing modality was 14-day extended lens wear for three months. The participants were assessed at lens dispensing, after one night, two weeks, one month and three months of extended wear and one month after study completion. (3) Results: There were no significant differences ($p > 0.05$) in ocular redness or palpebral roughness between Mel4 and control eyes at any of the study visits. There was no significant difference ($p > 0.05$) in corneal staining between Mel4 and control eyes. There were no significant differences in front surface wettability or deposits or back surface debris ($p > 0.05$). No statistically significant differences ($p > 0.05$) were found in comfort, dryness, CLDEQ-8 scores lens or edge awareness. There was no evidence for delayed reactions on the ocular surface after cessation of lens wear. (4) Conclusion: The novel MACLs showed similar comfort to control lenses and were biocompatible during extended wear. Thus, these lenses were compatible with the ocular surface.

Keywords: Mel4 peptide; antimicrobial contact lens; extended wear; biocompatibility; comfort; clinical trail

1. Introduction

Contact lens wear can be associated with inflammatory and infective responses, triggered by microbial colonisation of contact lenses. These are major concerns for contact lens wearers and practitioners. The development of contact lenses with antimicrobial activity may inhibit microbial adhesion and so reduce contact-lens-related inflammation and infection. Several antimicrobial contact lenses have been developed and tested in laboratory models. These include contact lenses containing silver [1–3], inhibitors of bacteria quorum-sensing systems [4], poly-epsilon lysine [5] and nitric-oxide-releasing lenses [6]. A cationic, peptide-coated (melimine) contact lens has also been developed that showed good antimicrobial activity in vitro [7], prevented bacterially-driven adverse events associated with contact lens wear in animal models [8,9] and was generally safe to wear in humans, although it was associated in some wearers with low levels of corneal staining [10]. Due to this latter issue, the cationic peptide melimine was shortened to make Mel4 [11].

Mel4, a small, cationic, antimicrobial peptide, has high antimicrobial activity against *Pseudomonas aeruginosa* and *Staphylococcus aureus* in solution and when immobilised on surfaces [12]. It has been successfully coated onto hydrogel and silicone hydrogel contact

lenses [12–14] and shown to be active against other bacteria such as *Stenotrophomonas maltophilia* and *Delftia acidovorans*. The Mel4-coated lenses were safe in a rabbit model of daily contralateral wear [13,14]. In addition, a phase I, human clinical trial showed no corneal fluorescein staining and no increase in ocular redness after one week of daily wear [13].

A phase II/III clinical trial on the Mel4-coated contact lenses was conducted at the LV Prasad Eye Institute in Hyderabad, India, and the biocompatibility of the lenses is addressed in the current manuscript. The main aim of this trial was to assess whether the Mel4-coated lenses could reduce the incidence of corneal inflammatory events during extended wear. These lenses resulted in a reduction in the incidence of corneal inflammatory events by 69% [15]. These Mel4-coated lenses had similar levels and types of microbes isolated from them and from eyes wearing them compared to the control lenses [16].

It is also valuable to investigate the biocompatibility and comfort of Mel4-coated contact lenses on the extended wear modality. Thus, the aim of the current study was to investigate the biocompatibility and comfort of Mel4-coated contact lenses during the phase II/III, extended wear, human clinical trial. The hypotheses of this study were that the Mel4-coated contact lenses are compatible and comfortable during extended wear.

2. Results

2.1. Assessment of Activity of Mel4-Coated Lenses Prior to Lens Wear

The data for the amount of Mel4 on contact lenses and the ability of lenses to inhibit the adhesion of *Pseudomonas aeruginosa* and *Staphylococcus aureus* have been previously published [15]. Briefly, randomly selected contact lenses from each batch that was produced were selected for measurement. These Mel4-coated lenses contained 62.6 ± 26.4 µg of amino acids per lens and significantly reduced the adhesion of *P. aeruginosa* and *S. aureus* by >1.8 log10 CFU/lens ($p < 0.001$) compared to control uncoated lenses [15]. This demonstrated that the participants in the trial were prescribed with active Mel4-coated contact lenses.

2.2. Subject Demographics

The demographic and biometric data for the subjects who were dispensed with study lenses are summarised in Table 1. Slightly more males (108/208; 52%) were enrolled and were dispensed with study lenses (93/176; 53%) than females. Additionally, a greater number of neophytes (160/208; 77%) were enrolled and were dispensed with study lenses (128/176; 73%). There was no difference in the refractive errors, keratometry and contact lens powers between the Mel4- and control-lens-wearing eyes in both enrolled ($p > 0.05$) and study-lens-dispensed subjects ($p > 0.05$).

Table 1. Demographic and biometric data of subjects dispensed with study lenses.

Demographic and Biometric Details	Subjects Dispensed with Study Lenses ($n = 176$)		p-Value
	Mel4-Lens-Wearing Eye	Control-Lens-Wearing Eye	
Age (years): Mean ± SD	22.6 ± 4.2		-
Range	18 to 42		
Gender (Male:Female)	93:83		-
Refractive error-Sphere (Ds) [1]: Mean ± SD; Range	−2.82 ± 1.44 −0.50 to −6.50	−2.80 ± 1.46 −0.50 to −6.50	0.528
Refractive error-Cylinder (Dc): Mean ± SD; Range	−0.25 ± 0.35 −0.25 to −1.50	−0.22 ± 0.35 −0.25 to −1.50	0.249
Keratometry-Flat (D): Mean ± SD; Range	43.02 ± 1.44 37.50 to 47.25	43.01 ± 1.46 37.50 to 47.25	0.663
Keratometry-Steep (D): Mean ± SD; Range	43.76 ± 1.58 38.75 to 48.50	43.71 ± 1.58 38.75 to 48.50	0.105
Contact lens wearer (Neophyte:Experienced lens wearer)	128:48		-
Contact lens base curve (8.3:8.7) mm	106:70		-
Contact lens power (Ds): Mean ± SD; Range	−2.84 ± 1.36 −1.00 to −6.00	−2.84 ± 1.39 −1.00 to −6.00	0.869

[1] D = diopter.

2.3. Clinical Lens Surface Characteristics

The lens surface characteristics of Mel4 and control contact lenses are presented in Table 2. There were no significant differences in front surface wettability between Mel4 and control lenses during all the visits ($p > 0.05$), with values ranging from 3.7 to 3.5 units. The front surface wetting for both Mel4 and control lenses decreased by 0.1–0.2 units over the course of the study, and this was significant ($p = 0.001$). There were no significant differences seen either in front surface deposits ($p > 0.05$) or back surface debris ($p > 0.05$) across all the study visits between both lens types. The front surface deposits for both lens types increased over the course of the study by between 0.1 and 0.5 units ($p = 0.001$), and the back surface debris significantly increased over the course of the study by between 0.1 and 0.3 units ($p = 0.001$).

Table 2. Surface characteristics of Mel4 and control contact lenses at various study visits.

Variables (Range, Incremental Steps)	Visits with Lens *	Number of Samples	Mel4 Lens (Mean ± SD)	Control Lens (Mean ± SD)	Linear Mixed Model Lens (Mel4 vs. Control)	Visit	Lens vs. Visit
Front surface wetting (0–4, 0.1)	Lens Dispensing	176	3.7 ± 0.6	3.7 ± 0.6	0.593	0.001	0.513
	1N	165	3.6 ± 0.6	3.7 ± 0.5			
	2W	153	3.6 ± 0.3	3.6 ± 0.3			
	1M	144	3.6 ± 0.3	3.5 ± 0.3			
	3M	128	3.6 ± 0.3	3.5 ± 0.3			
Front surface deposits (0–4, 0.1)	Lens Dispensing	176	0.2 ± 0.3	0.2 ± 0.3	0.896	0.001	0.996
	1N	165	0.3 ± 0.4	0.3 ± 0.4			
	2W	153	0.6 ± 0.5	0.6 ± 0.5			
	1M	144	0.6 ± 0.6	0.6 ± 0.6			
	3M	128	0.7 ± 0.6	0.7 ± 0.6			
Back surface debris (0–4, 0.1)	Lens Dispensing	176	0.1 ± 0.2	0.1 ± 0.2	0.715	0.001	0.857
	1N	165	0.2 ± 0.2	0.2 ± 0.2			
	2W	153	0.3 ± 0.4	0.3 ± 0.4			
	1M	144	0.4 ± 0.5	0.4 ± 0.5			
	3M	128	0.4 ± 0.5	0.4 ± 0.5			

* 1N = 1 night of lens wear, 2W = 2 weeks of lens wear, 1M = 1 month on lens wear, 3M = 3 months of lens wear.

2.4. Lens Fit Characteristics

The lens fit characteristics of Mel4 and control contact lenses are presented in Table 3. There were no significant differences in any lens fit characteristics between Mel4 and control contact lenses over the course of the study. The average overall lens acceptance score for both the lens types at each visit was 3.8 which indicated good centration, complete coverage, acceptable tightness of the lens and adequate lens movement and lens lag. No lens was refitted during the study period because of any lens fit issues. There were small but statistically significant differences between the visits for primary gaze movement, lag, tightness and overall acceptance ($p < 0.05$). There was no significant difference between the visits for lens centration ($p \geq 0.05$). No mucin balls were seen with either of the lens types at any of the visits.

Table 3. Lens fit characteristics of Mel4 and control contact lenses at various study visits.

Variables (Range, Incremental Steps)	Visits with Lens *	Number of Samples	Mel4 Lens (Mean ± SD)/ (Median and Range)	Control Lens (Mean ± SD)/ (Median and Range)	Linear Mixed Model		
					Lens	Visit	Lens vs. Visit
Centration X-axis (−1 to +1, 0.1 mm)	Lens Dispensing	176	0 (−0.3–0.2)	0 (−0.3–0.2)	0.589	0.409	0.950
	1N	165	0 (−0.5–0.3)	0 (−0.5–0.3)			
	2W	153	0 (−0.3–0.2)	0 (−0.3–0.2)			
	1M	144	0 (−0.5–0.2)	0 (−0.5–0.0)			
	3M	128	0 (−0.3–0.2)	0 (−0.3–0.2)			
Centration Y-axis (−1 to +1, 0.1 mm)	Lens Dispensing	176	0 (−0.3–0.5)	0 (−0.3–0.3)	0.595	0.118	0.266
	1N	165	0 (−0.4–0.5)	0 (−0.3–0.5)			
	2W	153	0 (−0.3–0.5)	0 (−0.3–0.5)			
	1M	144	0 (−0.4–0.4)	0 (−0.3–0.4)			
	3M	128	0 (−0.2–0.3)	0 (−0.2–0.3)			
Primary gaze movement (0–10, 0.1)	Lens Dispensing	176	0.4 ± 0.1	0.4 ± 0.1	0.988	0.049	0.752
	1N	165	0.4 ± 0.1	0.4 ± 0.1			
	2W	153	0.4 ± 0.1	0.4 ± 0.1			
	1M	144	0.4 ± 0.1	0.4 ± 0.1			
	3M	128	0.4 ± 0.1	0.4 ± 0.1			
Primary gaze lag (0–10, 0.1)	Lens Dispensing	176	0.1 ± 0.1	0.1 ± 0.1	1.000	0.005	-
	1N	165	0.1 ± 0.1	0.1 ± 0.1			
	2W	153	0.1 ± 0.1	0.1 ± 0.1			
	1M	144	0.2 ± 0.1	0.2 ± 0.1			
	3M	128	0.2 ± 0.1	0.2 ± 0.1			
Tightness (0–100, 1)%	Lens Dispensing	176	41 ± 3	41 ± 3	0.817	0.001	0.968
	1N	165	41 ± 3	41 ± 3			
	2W	153	42 ± 3	42 ± 3			
	1M	144	41 ± 3	41 ± 3			
	3M	128	42 ± 3	42 ± 3			
Overall acceptance (0–4, 0.1)	Lens Dispensing	176	3.8 ± 0.1	3.8 ± 0.1	0.410	0.001	0.380
	1N	165	3.8 ± 0.1	3.8 ± 0.1			
	2W	153	3.8 ± 0.1	3.8 ± 0.1			
	1M	144	3.8 ± 0.1	3.8 ± 0.1			
	3M	128	3.8 ± 0.1	3.8 ± 0.1			

* 1N = 1 night of lens wear, 2W = 2 weeks of lens wear, 1M = 1 month on lens wear, 3M = 3 months of lens wear.

2.5. Ocular Physiology

The conjunctival redness and roughness of the Mel4 and control lens wearing eye at the 1N, 2W, 1M and 3M study visits are presented in Table 4. There was no significant difference in bulbar redness ($p > 0.7$), limbal redness ($p > 0.9$), palpebral redness ($p > 0.6$) or palpebral roughness ($p > 0.3$) between Mel4- and control-contact-lens-wearing eyes in any of the study visits. All of these variables slightly but significantly ($p = 0.001$) increased by between 0.1 and 0.2 units over the course of the study. Similarly, there were no significant differences in lens-induced conjunctival staining ($p = 1.0$) or indentation ($p > 0.1$) between the Mel4-lens-wearing eye and control-lens-wearing eye (Table 4). There were no significant differences ($p > 0.35$) in central, nasal, temporal or superior corneal staining (extent, depth or type) between Mel4- and control-contact-lens-wearing eyes (Table 5). None of these corneal-staining characteristics changed during the study ($p \geq 0.2$).

Table 4. Conjunctival responses during contact lens wear.

Variables (Range, Incremental Steps)	Visits with Lens *	Number of Samples	Mel4 Lens (Mean ± SD)	Control Lens (Mean ± SD)	Linear Mixed Model		
					Lens (Mel4 vs. Control)	Visit	Lens vs. Visit
Bulbar Redness (0–4, 0.1)	1N	167	1.5 ± 0.2	1.5 ± 0.2	0.758	0.001	0.135
	2W	153	1.5 ± 0.2	1.5 ± 0.2			
	1M	144	1.6 ± 0.2	1.6 ± 0.2			
	3M	129	1.6 ± 0.2	1.6 ± 0.2			
Limbal Redness (0–4, 0.1)	1N	167	1.2 ± 0.3	1.2 ± 0.3	0.961	0.001	0.660
	2W	153	1.2 ± 0.2	1.2 ± 0.2			
	1M	144	1.3 ± 0.3	1.3 ± 0.3			
	3M	129	1.4 ± 0.2	1.4 ± 0.2			
Palpebral Redness (0–4, 0.1)	1N	167	1.5 ± 0.2	1.6 ± 0.2	0.610	0.001	0.053
	2W	153	1.6 ± 0.3	1.6 ± 0.3			
	1M	144	1.6 ± 0.3	1.6 ± 0.3			
	3M	129	1.7 ± 0.3	1.7 ± 0.3			
Palpebral Roughness (0–4, 0.1)	1N	167	1.2 ± 0.3	1.3 ± 0.3	0.388	0.001	0.574
	2W	153	1.3 ± 0.3	1.3 ± 0.3			
	1M	144	1.3 ± 0.3	1.3 ± 0.3			
	3M	129	1.4 ± 0.3	1.4 ± 0.3			
Lens Induced Conjunctival Staining (0–4, 0.1)	1N	167	0.2 ± 0.2	0.2 ± 0.2	1.000	0.001	1.000
	2W	153	0.2 ± 0.3	0.2 ± 0.2			
	1M	144	0.3 ± 0.3	0.3 ± 0.3			
	3M	129	0.3 ± 0.3	0.3 ± 0.3			
Lens Induced Conjunctival Indentation (0–4, 0.1)	1N	167	0.1 ± 0.1	0.1 ± 0.1	0.112	0.001	0.041
	2W	153	0.1 ± 0.2	0.1 ± 0.2			
	1M	144	0.1 ± 0.2	0.1 ± 0.2			
	3M	129	0.1 ± 0.2	0.1 ± 0.2			

* 1N = 1 night of lens wear, 2W = 2 weeks of lens wear, 1M = 1 month on lens wear, 3M = 3 months of lens wear.

Table 5. Corneal staining during contact lens wear.

Variables (Type; Range, Incremental Steps)	Visits with Lens *	Number of Samples	Mel4 Lens (Median and Range)	Control Lens (Median and Range)	Linear Mixed Model		
					Lens (Mel4 vs. Control)	Visit	Lens vs. Visit
Centre (Extent; 0–4, 1)	1N	162	0 (0–0)	0 (0–0)	0.674	0.200	0.632
	2W	153	0 (0–1)	0 (0–1)			
	1M	144	0 (0–0)	0 (0–1)			
	3M	129	0 (0–1)	0 (0–1)			
Centre (Depth; 0–4, 1)	1N	162	0 (0–0)	0 (0–0)	0.674	0.200	0.632
	2W	153	0 (0–1)	0 (0–1)			
	1M	144	0 (0–0)	0 (0–1)			
	3M	129	0 (0–1)	0 (0–1)			
Centre (Type; 0–4, 0.5)	1N	162 *	0 (0–0)	0 (0–0)	0.674	0.200	0.632
	2W	153	0 (0–1)	0 (0–1)			
	1M	144	0 (0–0)	0 (0–1)			
	3M	129	0 (0–1)	0 (0–1)			
Nasal (Extent; 0–4, 1)	1N	162	0 (0–1)	0 (0–1)	0.670	0.426	0.262
	2W	153	0 (0–1)	0 (0–1)			
	1M	144	0 (0–1)	0 (0–2)			
	3M	129	0 (0–0)	0 (0–1)			
Nasal (Depth; 0–4, 1)	1N	162	0 (0–1)	0 (0–1)	0.869	0.421	0.254
	2W	153	0 (0–1)	0 (0–1)			
	1M	144	0 (0–1)	0 (0–2)			
	3M	129	0 (0–0)	0 (0–1)			

Table 5. *Cont.*

Variables (Type; Range, Incremental Steps)	Visits with Lens *	Number of Samples	Mel4 Lens (Median and Range)	Control Lens (Median and Range)	Linear Mixed Model		
					Lens (Mel4 vs. Control)	Visit	Lens vs. Visit
Nasal (Type; 0–4, 0.5)	1N	162	0 (0–1)	0 (0–1.5)	0.586	0.386	0.342
	2W	153	0 (0–1)	0 (0–1)			
	1M	144	0 (0–1)	0 (0–2)			
	3M	129	0 (0–0)	0 (0–1)			
Temporal (Extent; 0–4, 1)	1N	162	0 (0–1)	0 (0–1)	0.856	0.437	0.457
	2W	153	0 (0–1)	0 (0–1)			
	1M	144	0 (0–0)	0 (0–1)			
	3M	129	0 (0–1)	0 (0–0)			
Temporal (Depth; 0–4, 1)	1N	162	0 (0–1)	0 (0–1)	0.856	0.437	0.457
	2W	153	0 (0–1)	0 (0–1)			
	1M	144	0 (0–0)	0 (0–1)			
	3M	129	0 (0–1)	0 (0–0)			
Temporal (Type; 0–4, 0.5)	1N	162	0 (0–1)	0 (0–1)	0.780	0.340	0.498
	2W	153	0 (0–1)	0 (0–1)			
	1M	144	0 (0–0)	0 (0–0.5)			
	3M	129	0 (0–1)	0 (0–0)			
Superior (Extent; 0–4, 1)	1N	162	0 (0–0)	0 (0–1)	0.368	0.242	0.362
	2W	153	0 (0–1)	0 (0–1)			
	1M	144	0 (0–1)	0 (0–2)			
	3M	129	0 (0–1)	0 (0–1)			
Superior (Depth; 0–4, 1)	1N	162	0 (0–0)	0 (0–1)	0.368	0.242	0.362
	2W	153	0 (0–1)	0 (0–1)			
	1M	144	0 (0–1)	0 (0–2)			
	3M	129	0 (0–1)	0 (0–1)			
Superior (Type; 0–4, 0.5)	1N	162	0 (0–0)	0 (0–1)	0.649	0.151	0.258
	2W	153	0 (0–1)	0 (0–0.5)			
	1M	144	0 (0–1)	0 (0–1.5)			
	3M	129	0 (0–1)	0 (0–1)			
Inferior (Extent; 0–4, 1)	1N	162	0 (0–1)	0 (0–1)	0.119	0.238	0.337
	2W	153	0 (0–1)	0 (0–2)			
	1M	144	0 (0–1)	0 (0–2)			
	3M	129	0 (0–2)	0 (0–2)			
Inferior (Depth; 0–4, 1)	1N	162	0 (0–1)	0 (0–1)	0.119	0.238	0.337
	2W	153	0 (0–1)	0 (0–2)			
	1M	144	0 (0–1)	0 (0–2)			
	3M	129	0 (0–2)	0 (0–2)			
Inferior (Type; 0–4, 0.5)	1N	162	0 (0–1)	0 (0–1)	0.119	0.238	0.337
	2W	153	0 (0–1)	0 (0–2)			
	1M	144	0 (0–1)	0 (0–2)			
	3M	129	0 (0–2)	0 (0–2)			

* 1N = 1 night of lens wear, 2W = 2 weeks of lens wear, 1M = 1 month on lens wear, 3M = 3 months of lens wear.

2.6. Subjective Ratings

The subjective ratings of the comfort at each visit are presented in Table 6. There was no significant difference ($p > 0.1$) in the subjective ratings of overall comfort, dryness or lens edge awareness between Mel4- and control-lens-wearing eyes. Overall comfort significantly ($p = 0.001$) decreased for both lens types from the 1N to the 3M visits, dropping by 3–4 points, as did overall dryness ($p = 0.001$) which increased by 5 points.

Table 6. Ocular comfort responses during contact lens wear.

Variables (Range, Incremental Steps)	Visits with Lens *	Number of Samples	Mel4 Lens (Mean ± SD)	Control Lens (Mean ± SD)	Linear Mixed Model Lens (Mel4 vs. Control)	Visit	Lens vs. Visit
Overall comfort (1–100, 1)	1N	167	92 ± 8	92 ± 8	0.770	0.001	0.556
	2W	153	91 ± 7	91 ± 8			
	1M	144	90 ± 7	91 ± 9			
	3M	129	88 ± 8	89 ± 7			
Overall dryness (1–100, 1)	1N	167	8 ± 5	8 ± 5	0.789	0.001	0.742
	2W	153	11 ± 11	11 ± 11			
	1M	144	11 ± 11	12 ± 12			
	3M	129	13 ± 11	13 ± 11			
Edge awareness (0–10, 1)	1N	165	1 (1–2)	1 (1–2)	0.135	0.077	0.969
	2W	153	1 (1–4)	1 (1–5)			
	1M	144	1 (1–3)	1 (1–3)			
	3M	128	1 (1–3)	1 (1–3)			
Lens awareness (0–10, 1)	1N	165	1 (1–2)	1 (1–2)	0.139	0.471	0.858
	2W	153	1 (1–4)	1 (1–4)			
	1M	144	1 (1–3)	1 (1–3)			
	3M	128	1 (1–3)	1 (1–3)			

* 1N = 1 night of lens wear, 2W = 2 weeks of lens wear, 1M = 1 month on lens wear, 3M = 3 months of lens wear.

All subjects were asked to score the modified CLDEQ-8 questionnaires at the one-month and three-month visit. One hundred and thirty-nine subjects scored the questionnaires at the first month's visit and 126 subjects scored the questionnaires at the three-month visit. There were no significant differences in CLDEQ-8 scores between Mel4- and control-lens-wearing eyes at either the one- ($p = 0.5$) or three-month visit ($p = 0.9$) (Figure 1).

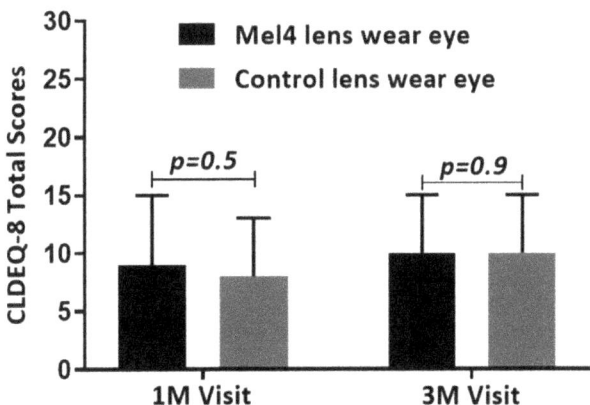

Figure 1. Comfort of contact lenses during wear measured with the modified CLDEQ-8.

2.7. Ocular Responses One Month after Cessation of Lens Wear

The ocular responses at the four-month visit, i.e., one month after cessation of lens wear, are provided in Supplementary Table S2. There were no statistical differences between the Mel4-lens-wearing eyes and the control-lens-wearing eyes for any of the clinical or subjective variables. There were some small but significant differences between the study visits for conjunctival bulbar, limbal and palpebral redness ($p = 0.001$), which had increased by 0.1 to 0.2 units, and palpebral roughness ($p = 0.001$), which had increased by 0.1 unit. There was also a very small but significant change in inferior corneal staining (extent, depth and type; $p \leq 0.016$), which had increased by less than a unit interval. Overall, ocular comfort was slightly but significantly ($p = 0.001$) decreased at the four-month visit by 3 units. As each of these differences were not different between lenses and there were no lens/visit interactions in the statistical analyses, these changes were likely the result of lens wear rather than the result of wearing either Mel4-coated or control lenses.

2.8. Ocular Responses of Participants Who Dropped out of Lens Wear during the Study Compared to Those Who Completed the Study

The ocular and subjective responses of the participants who dropped out of lens wear during the study compared to those who completed the study at different study visits are provided in Supplementary Table S3. The responses collected from four participants at the 1N visit that had dropped out by the 2W visit were compared to the participants who remained wearing lenses at the 2W visit (153). Similarly, two participants at the 2W visit that had dropped out by the 1M visit and their responses were compared to the participants who remained wearing lenses at the 1M visit (144), and two participants at the 1M visit that had dropped out by the 3M visit were compared to the participants who remained wearing lenses at the 3M visit (129). There were no clinical differences in ocular and subjective responses between dropouts and those who remained in Mel4 or control lenses. The lens surface characteristics and lens fit characteristics of the participants who dropped out of lens wearing during the study compared to those who completed the study at different study visits are provided in Supplementary Table S4. There were no clinical differences in lens surface characteristics and lens fit characteristics between dropouts and those who remained in Mel4 or control lenses.

3. Discussion

This study investigated the biocompatibility and comfort of Mel4-coated contact lenses in a human, three-month, extended wear clinical trial. Overall, in comparison to the eyes wearing control, uncoated, etafilcon A lenses, Mel4-coated lenses had no worse effect on the biocompatibility and comfort during lens wear than the uncoated lenses. This indicates that Mel4-coated lenses are not prone to forming more deposits, wet as well as control lenses, do not affect lens parameters that could influence centration, movement or tightness and do not induce inflammation as measured by changes in redness or conjunctival or corneal staining or comfort during lens wear. This is in agreement with a one-week study of daily wear with Mel4-coated lenses [13] and is similar to results reported for other antimicrobial lenses that had been made using a fimbrolide [4] or another cationic, antimicrobial peptide (a forerunner of Mel4), melimine [10]. Wearing Mel4-coated contact lenses over a three-month period did not result in any delayed ocular reactions as the data from the 4M visit, at which time the participants had been not wearing Mel4-coated lenses for one month (but had returned to their habitual method for correcting their refractive error), were not different between the Mel4- or control-lens-wearing eyes.

This study found no significant fluorescein staining (extent, depth and type) of the cornea following lens wear of Mel4-coated lenses, which is similar to the study on Mel4-coated lenses worn on a daily-wear basis for a week [13] and fimbrolide-coated lenses after 20 to 22 h of lens wear [4]. However, this does contrast with a study of melimine-coated contact lenses which were associated with corneal staining [10]. The current study confirms that the change in amino acid sequence between melimine (TLISWIKNKRKQRPRVSR-RRRRGGRRRR) and Mel4 (KNKRKRRRRRRGGRRRR) eliminated the corneal punctate staining that occurred with melimine-coated lenses. The change from melimine to Mel4 removed several amino acids. Tryptophan (W) is present in the amino acid sequence of melimine at position five. Also present in melimine, are other hydrophobic amino acids such as isoleucine (I, two residues at positions 3 and 6), glycine (G, two residues at positions 24 and 25), valine (V, one residue at position 16), leucine (L, one residue at position 2) and proline (P, one residue at position 14). None of these amino acids are in Mel4. It is possible that one or more of these might have been involved in interactions with human corneal cells. The amino acid tryptophan (W) is often present in proteins that reside within membranes [17,18] which indicates its potential to interact with membranes. Additionally, tryptophan in arginine-rich peptides facilitates the translocation of these peptides through membranes [19].

Similar to melimine coating [7], the Mel4 coating (22.7° ± 5.0) has previously been shown to significantly improve the wettability of etafilcon A lenses (69.3° ± 14.6) when measured using the advancing contact angle technique in the laboratory [13]. This reinforces previous findings that wettability measured in the laboratory does not translate to improved comfort responses [20] as there was no difference in comfort between Mel4-coated and control lenses. Fimbrolide-coated lenses [4] have been associated with increased dryness, lens edge and lens awareness and were slightly less comfortable to wear. In the current study, also there was no significant difference in dryness, lens edge and lens awareness between Mel4-coated and control lenses, similar to the study of Mel4-coated lenses [13] used on a daily-wear basis for a week or melimine-coated lenses [10] when worn for a day. The reason for the improved comfort response with Mel4-coated lenses compared to fimbrolide lenses may be due to the different antimicrobial compounds themselves, the different chemistries used to attach fimbrolide and Mel4 to lenses or the different lenses used (hydrogel etafilcon A vs. silicone hydrogel lotrafilcon A).

To understand whether dropout from this clinical trial at any stage was associated with differences in the ocular surface or lens characteristics of the people who dropped out compared to those that completed the trial, variables of those who dropped out were compared to those who remained in the trial at each visit and at the final 3-month visit. This demonstrated that the people who dropped out did not have any significant differences in ocular surface responses, lens characteristics or comfort. This further reinforces the biocompatibility of Mel4 during lens wear.

4. Materials and Methods
4.1. Study Design and Participants

The study design, inclusion and exclusion criteria were identical to a previous study [16]. Briefly, a total of 176 participants who met the inclusion/exclusion criteria were dispensed with study lenses. Participants were randomly assigned to wear a Mel4 antimicrobial contact lens (MACL) in one eye and a control lens (uncoated etafilcon A) in the contralateral eye. All the participants were instructed to replace lenses every two weeks during three months of extended wear. To reduce the possibility of participants mixing right- and left-eye lenses, the right-eye and left-eye lens vials were affixed with green and white labels, respectively. If the participants needed to remove their lenses temporarily, they were given Biotrue contact lens care solution (Bausch and Lomb, Rochester, NY, USA) and a case only for temporary storage.

4.2. Production and Quantification of Mel4 Peptide Attached to Contact Lenses

Etafilcon A contact lenses (Acuvue2®, Johnson and Johnson Vision Care Inc., Jacksonville, FL, USA) were used for this study. Mel4 peptide (amino acid sequence: KNKRKR-RRRRRGGRRRR; American Peptide Company, Sunnyvale, CA, USA) was synthesised by conventional solid-phase peptide synthesis with >95% purity. The procedure for covalently attaching Mel4 to contact lenses has been reported elsewhere [10,13,16]. Control, uncoated (etafilcon A) lenses were removed from their packs, washed and autoclaved prior to use. Mel4-coated lenses for the clinical trial were produced in different batches. After the production of each batch and prior to the lens dispensing visit, two Mel4-coated lenses from each batch were assessed by amino acid analysis to confirm the presence and amount of peptide on to the lens surface [12,13,21]. The sum of all the amino acids derived from each contact lens was regarded as the total amount of peptide attachment to a contact lens. Similarly, two Mel4-coated and control lenses from each batch were assessed for adhesion of *P. aeruginosa* (ATCC 27853) and *S. aureus* (L2260/15). The bacterial adhesion protocol has been reported previously [7,12].

4.3. Clinical Procedures

A total of seven visits were undertaken; baseline (visit 1), lens dispensing (visit 2), after one night (visit 3), 2 weeks, (visit 4), 1 month (visit 5) and 3 months of lens wear (visit 6), followed by a 1-month follow-up visit after study lens discontinuation (visit 7). The seventh follow-up visit at the end of 3 months' extended wear included no assigned contact lens wear, and the subjects were free to wear their glasses if desired to test for any delayed responses to the investigational product. At each scheduled visit, the ocular characteristics and subjective responses of each participant were assessed. Slit-lamp biomicroscopy was performed for anterior eye assessment including contact lens fit, lens surface characteristics, ocular redness (bulbar, limbal and palpebral), palpebral roughness, conjunctival and corneal staining. Contact lens fitting (assessed using the push-up test) and contact lens deposits during wear were measured according to previously described methods [22,23] All of the clinical grading was conducted using the Cornea and Contact Lens Research Unit grading scales [24] (0 to 4 units) interpolated into 0.1 increments, except for corneal staining which was graded in 1.0 steps for extent and depth and 0.5 steps for type. Concordance training for the optometrists was conducted before study commencement, and concordance was measured every 6 months during the study. All optometrists were allowed to examine study participants if they scored more than 70% concordance for each grading scale, and they were retrained if concordance dropped below this level. All optometrists were masked to which eye of each participant was wearing which contact lenses. The CLDEQ-8 questionnaire was modified for monocular lens wear (see Supplementary Table S1).

4.4. Statistical Analysis

Data were analysed using Microsoft® Office Excel®, Graph Pad Prism 7.02 (Graph Pad Software Inc., San Diego, CA, USA) and IBM SPSS (Package for the Social Sciences software) for Windows software v24.0 (SPSS, Inc., Chicago, IL, USA). According to central limit theorem, [25] this study, with its large sample collection of ordinal variables, can be assumed to have a normal distribution; thus, the means of the variables were reported with standard deviation. Non-ordinal variables were described as median and range. The comparisons of clinical and lens variables at each visit between Mel4- and control-lens-wearing eyes were examined using linear mixed models (LMM). As some subjects dropped out of lens wear during the study and there were some other cases of missing data, the sample size varied among the study visits and so LMM analyses were conducted. Post hoc *t*-tests with Bonferroni correction were used to assess significant changes over time and between both lens-wear types. For the subjects who dropped out of the study, where data had been collected, this was compared to the subjects who remained in the study to determine whether changes to ocular surface physiology or comfort during wear may have influenced the decision to drop out of the study. For all tests, the level of statistical significance was maintained at $p < 0.05$.

5. Conclusions

In conclusion, wearing Mel4-coated contact lenses did not affect the ocular responses, contact lens deposition or other characteristics or comfort during lens wear. These data, combined with the fact that the Mel4-coated lenses did not affect the normal ocular microbiota during wear [16] and could reduce the number of corneal infiltrative events during extended contact lens wear [15], indicate that Mel4 coatings are biocompatible and useful to control microbially-driven adverse events.

Supplementary Materials: The following supporting information can be downloaded at: https://www.mdpi.com/article/10.3390/antibiotics11010058/s1, Table S1: Modified CLDEQ-8 questionnaire for monocular lens wear, Table S2: Ocular and subjective responses at baseline visit prior to lens wear and at the 4M study visit, one month after cessation of contact lens wear, Table S3: Ocular and subjective responses at dropouts and completed visits, Table S4: Lens surface characteristics and fit characteristics of Mel4 and control contact lenses at dropouts and completed visits.

Author Contributions: Conceptualization, D.D. and M.D.P.W.; methodology, P.K., D.D., N.K., P.K.V., S.S., F.S. and M.D.P.W.; formal analysis, P.K., D.D. and M.D.P.W.; investigation, P.K., D.D., N.K., P.K.V., S.S., F.S. and M.D.P.W.; resources, F.S. and M.D.P.W.; writing—original draft preparation, P.K.; writing—review and editing, P.K., D.D., N.K., P.K.V., S.S., F.S. and M.D.P.W.; supervision, D.D., N.K. and M.D.P.W.; project administration, D.D. and N.K.; funding acquisition, F.S. and M.D.P.W. All authors have read and agreed to the published version of the manuscript.

Funding: This research was funded by the National Health and Medical Research Council, grant number APP1076206, and a grant from CooperVision, Pleasanton, CA, USA.

Institutional Review Board Statement: This study received approval from the Human Research Ethics Committee of the University of New South Wales (ethics number HC15436) and the Institutional Ethics Committee of Hyderabad Eye Research Foundation (ethics number LEC 05_15-057). The study was registered at the Clinical Trials Registry—India (CTRI; trial ID: CTRI/2015/10/006,327) following approval from the Central Drugs Standard Control Organization (CDSCO) and Drug Controller General of India (DCGI:4-MD/CT-148/2015- DC), India.

Informed Consent Statement: This research was conducted in accordance with the Declaration of Helsinki (2013). Written informed consent was obtained from all participants after an explanation of the purpose of the study before any study procedures were performed.

Data Availability Statement: Data are available upon request from the corresponding author.

Acknowledgments: All etafilcon A contact lenses were supplied by Johnson and Johnson Vision Care, Jacksonville, FL, USA.

Conflicts of Interest: The authors declare no conflict of interest.

References

1. Fazly Bazzaz, B.S.; Khameneh, B.; Jalili-Behabadi, M.M.; Malaekeh-Nikouei, B.; Mohajeri, S.A. Preparation, characterization and antimicrobial study of a hydrogel (soft contact lens) material impregnated with silver nanoparticles. *Cont. Lens Anterior Eye* **2014**, *37*, 149–152. [CrossRef] [PubMed]
2. No, J.W.; Kim, D.H.; Lee, M.J.; Kim, D.H.; Kim, T.H.; Sung, A.Y. Preparation and characterization of ophthalmic lens materials containing titanium silicon oxide and silver nanoparticles. *J. Nanosci. Nanotechnol.* **2015**, *15*, 8016–8022. [CrossRef] [PubMed]
3. Liu, X.; Chen, J.; Qu, C.; Bo, G.; Jiang, L.; Zhao, H.; Zhang, J.; Lin, Y.; Hua, Y.; Yang, P.; et al. A mussel-inspired facile method to prepare multilayer-AgNP-loaded contact lens for early treatment of bacterial and fungal keratitis. *ACS Biomater. Sci. Eng.* **2018**, *4*, 1568–1579. [CrossRef] [PubMed]
4. Zhu, H.; Kumar, A.; Ozkan, J.; Bandara, R.; Ding, A.; Perera, I.; Steinberg, P.; Kumar, N.; Lao, W.; Griesser, S.S.; et al. Fimbrolide-coated antimicrobial lenses: Their in vitro and in vivo effects. *Optom. Vis. Sci.* **2008**, *85*, 292–300. [CrossRef]
5. Gallagher, A.G.; Alorabi, J.A.; Wellings, D.A.; Lace, R.; Horsburgh, M.J.; Williams, R.L. A novel peptide hydrogel for an antimicrobial bandage contact lens. *Adv. Healthc. Mater.* **2016**, *5*, 2013–2018. [CrossRef] [PubMed]
6. Aveyard, J.; Deller, R.C.; Lace, R.; Williams, R.L.; Kaye, S.B.; Kolegraff, K.N.; Curran, J.M.; D'Sa, R.A. Antimicrobial nitric oxide releasing contact lens gels for the treatment of microbial keratitis. *ACS Appl. Mater. Interfaces* **2019**, *11*, 37491–37501. [CrossRef]
7. Dutta, D.; Cole, N.; Kumar, N.; Willcox, M.D. Broad spectrum antimicrobial activity of melimine covalently bound to contact lenses. *Invest. Ophthalmol. Vis. Sci.* **2013**, *54*, 175–182. [CrossRef]
8. Dutta, D.; Vijay, A.K.; Kumar, N.; Willcox, M.D. Melimine-coated antimicrobial contact lenses reduce microbial keratitis in an animal model. *Invest. Ophthalmol. Vis. Sci.* **2016**, *57*, 5616–5624. [CrossRef]
9. Cole, N.; Hume, E.B.; Vijay, A.K.; Sankaridurg, P.; Kumar, N.; Willcox, M.D. In vivo performance of melimine as an antimicrobial coating for contact lenses in models of CLARE and CLPU. *Invest. Ophthalmol. Vis. Sci.* **2010**, *51*, 390–395. [CrossRef] [PubMed]
10. Dutta, D.; Ozkan, J.; Willcox, M.D. Biocompatibility of antimicrobial melimine lenses: Rabbit and human studies. *Optom. Vis. Sci.* **2014**, *91*, 570–581. [CrossRef]
11. Willcox, M.D.P.; Chen, R.; Kalaiselvan, P.; Yasir, M.; Rasul, R.; Kumar, N.; Dutta, D. The development of an antimicrobial contact lens—From the laboratory to the clinic. *Curr. Protein Pept. Sci.* **2020**, *21*, 357–368. [CrossRef]
12. Dutta, D.; Kumar, N.; Willcox, M.D.P. Antimicrobial activity of four cationic peptides immobilised to poly-hydroxyethylmethacrylate. *Biofouling* **2016**, *32*, 429–438. [CrossRef]
13. Dutta, D.; Zhao, T.; Cheah, K.B.; Holmlund, L.; Willcox, M.D.P. Activity of a melimine derived peptide Mel4 against *Stenotrophomonas, Delftia, Elizabethkingia, Burkholderia* and biocompatibility as a contact lens coating. *Cont. Lens Anterior Eye* **2017**, *40*, 175–183. [CrossRef]
14. Dutta, D.; Kamphuis, B.; Ozcelik, B.; Thissen, H.; Pinarbasi, R.; Kumar, N.; Willcox, M.D.P. Development of silicone hydrogel antimicrobial contact lenses with Mel4 peptide coating. *Optom. Vis. Sci.* **2018**, *95*, 937–946. [CrossRef] [PubMed]

15. Kalaiselvan, P.; Konda, N.; Pampi, N.; Vaddavalli, P.K.; Sharma, S.; Stapleton, F.; Kumar, N.; Willcox, M.D.P.; Dutta, D. Effect of antimicrobial contact lenses on corneal infiltrative events: A randomized clinical trial. *Transl. Vis. Sci. Technol.* **2021**, *10*, 32. [CrossRef] [PubMed]
16. Kalaiselvan, P.; Dutta, D.; Bhombal, F.; Konda, N.; Vaddavalli, P.K.; Sharma, S.; Stapleton, F.; Willcox, M.D.P. Ocular microbiota and lens contamination following Mel4 peptide-coated antimicrobial contact lens (MACL) extended wear. *Cont. Lens Anterior Eye* **2021**, 101431. [CrossRef] [PubMed]
17. Landolt-Marticorena, C.; Williams, K.A.; Deber, C.M.; Reithmeier, R.A. Non-random distribution of amino acids in the transmembrane segments of human type I single span membrane proteins. *J. Mol. Biol.* **1993**, *229*, 602–608. [CrossRef]
18. Reithmeier, R.A. Characterization and modeling of membrane proteins using sequence analysis. *Curr. Opin. Struct. Biol.* **1995**, *5*, 491–500. [CrossRef]
19. Christiaens, B.; Symoens, S.; Verheyden, S.; Engelborghs, Y.; Joliot, A.; Prochiantz, A.; Vandekerckhove, J.; Rosseneu, M.; Vanloo, B. Tryptophan fluorescence study of the interaction of penetratin peptides with model membranes. *Eur. J. Biochem.* **2002**, *269*, 2918–2926. [CrossRef]
20. Willcox, M.; Keir, N.; Maseedupally, V.; Masoudi, S.; McDermott, A.; Mobeen, R.; Purslow, C.; Santodomingo-Rubido, J.; Tavazzi, S.; Zeri, F.; et al. CLEAR—Contact lens wettability, cleaning, disinfection and interactions with tears. *Cont. Lens Anterior Eye* **2021**, *44*, 157–191. [CrossRef]
21. Kaspar, H.; Dettmer, K.; Gronwald, W.; Oefner, P.J. Advances in amino acid analysis. *Anal. Bioanal. Chem.* **2009**, *393*, 445–452. [CrossRef] [PubMed]
22. Young, G. Evaluation of soft contact lens fitting characteristics. *Optom. Vis. Sci.* **1996**, *73*, 247–254. [CrossRef]
23. Stern, J.; Wong, R.; Naduvilath, T.J.; Stretton, S.; Holden, B.A.; Sweeney, D.F. Comparison of the performance of 6- or 30-night extended wear schedules with silicone hydrogel lenses over 3 years. *Optom. Vis. Sci.* **2004**, *81*, 398–406. [CrossRef] [PubMed]
24. Terry, R.L.; Schnider, C.M.; Holden, B.A.; Cornish, R.; Grant, T.; Sweeney, D.; La Hood, D.; Back, A. CCLRU standards for success of daily and extended wear contact lenses. *Optom. Vis. Sci.* **1993**, *70*, 234–243. [CrossRef]
25. Kwak, S.G.; Kim, J.H. Central limit theorem: The cornerstone of modern statistics. *Korean J. Anesthesiol.* **2017**, *70*, 144–156. [CrossRef] [PubMed]

MDPI
St. Alban-Anlage 66
4052 Basel
Switzerland
Tel. +41 61 683 77 34
Fax +41 61 302 89 18
www.mdpi.com

Antibiotics Editorial Office
E-mail: antibiotics@mdpi.com
www.mdpi.com/journal/antibiotics

www.ingramcontent.com/pod-product-compliance
Lightning Source LLC
LaVergne TN
LVHW070623100526
838202LV00012B/707